Dietmar Pohl
with Carl Schulze

MODERN KNIVES IN COMBAT

Schiffer Military History
Atglen, PA

Photos: Carl Schulze (all action photos unless stated otherwise), Dietmar Pohl, Knife Maker

Translated from the German by David Johnston.

Original book design by Caroline Wydeau and Andreas Schleicher,
Revised by Stephanie Daugherty.

Library of Congress Control Number: 2011922163

Printed in China.
ISBN: 978-0-7643-3766-6

This book was originally published in German under the title
Messer im Kampfeinsatz by Wieland Verlag GmbH

We are interested in hearing from authors with book ideas on related topics.

Published by Schiffer Publishing Ltd.
4880 Lower Valley Road
Atglen, PA 19310
Phone: (610) 593-1777
FAX: (610) 593-2002
E-mail: Info@schifferbooks.com.
Visit our web site at: www.schifferbooks.com
Please write for a free catalog.
This book may be purchased from the publisher.
Please include $5.00 postage.
Try your bookstore first.

In Europe, Schiffer books are distributed by:
Bushwood Books
6 Marksbury Avenue
Kew Gardens
Surrey TW9 4JF, England
Phone: 44 (0) 20 8392-8585
FAX: 44 (0) 20 8392-9876
E-mail: Info@bushwoodbooks.co.uk.
Visit our website at: www.bushwoodbooks.co.uk
Free postage in the UK. Europe: air mail at cost.
Try your bookstore first.

A fighting axe and a combat knife made by the American knife maker Daniel Winkler and developed by him for American special forces. On the right is a KM2K combat knife as used by the Bundeswehr, produced by the company Waffentechnik Borkott & Eickhorn.

CONTENTS

The knife as helper in daily life.

INTRODUCTION

Even in the days of modern warfare, knives and bayonets remain indispensible items of equipment for the soldier, whether as tools or edged weapons. This book graphically illustrates this. Over a period of about ten years, Carl Schulze photographed a range of knives used by soldiers in Afghanistan, Iraq and the Balkans.

On viewing the photos, it very quickly becomes apparent that, in addition to the knife or bayonet issued to them, most soldiers also carry a second, usually smaller, knife which meets their individual needs. Quite often the knives officially issued to them are simply unsuitable for everyday tasks as they were designed primarily as combat knives. The vast majority of the knives carried are used primarily as work knives, with which one can open a can or package or carry out other manual tasks.

When looking at the various knives and their manufacturers, one realizes that most are at the lower end of the price range. This is due in part to the fact that the soldiers must purchase these knives using their own resources and consequently have no interest in a major investment. On the other hand, many soldiers fear that their knives will be lost in action.

It is also noticeable that certain manufacturers are better represented than others. This is in part due to the ability of soldiers to purchase certain items of equipment at reasonable price sin military shops in the countries they deploy to. Those companies that offer a good assortment in these stores also do more business. Of course, here and there one also encounters more expensive knives, carried by knife enthusiasts, for example the hand made knives of Randall Knives or Vox Knives. These are a relatively small minority, however.

Makers of high-quality tactical combat knives like to advertise that their knives have proved themselves in the struggle against terrorism and were used by many members of special units. This cannot be confirmed either in Iraq or Afghanistan. On closer observation, one also sees that the choice of knife is influenced not only by rank and consequently rate of pay, but also by ethnic and social backgrounds. Once again this shows that the military forces are a reflection of the society they come from.

In addition to readily available knives, there are also specialized knives, some of which have a single function. Typical of these are the tiny folding knives used to free a wounded man from his equipment as quickly as possible. Axes and tomahawks have experienced a true resurgence, in part as tools for clearing obstacles, and also as weapons used in close combat.

Views on the use of bayonets have also undergone a fundamental change based on operational experience. While not long ago many "experts" considered the bayonet an obsolete relic from the time of the First World War, it has been shown that even today – in the times of asymmetric warfare – that fixed bayonets can be the soldier's last weapon in close combat. As well, the bayonet can serve as an extension of the arm when searching for concealed weapons, possibly guarded by booby traps.

The documentation at hand is unique in that, until now, there have been no photographs on this theme. Some were taken at the risk of the photographer's life. The variations in quality can be explained by the mixed use of film and modern digital photography.

FOLDING KNIVES

Using a map of Basra, a British officer of the 7th Armoured Brigade "Desert Rats" briefs an Iraqi comrade on their patrol route. The objective of joint patrols by British and Iraqi soldiers was to increase the civilian population's trust in the security forces. They also served to improve the training of the Iraqi soldiers and motivate their officers to conduct more patrols. Taken in the summer of 2008, this photo shows the British officer using his M16SF single-hand knife made by CRKT as a pointer.

Afghanistan: In addition to his Wiley-X protective glasses, this Norwegian ISAF soldier from Camp Invicta in Kabul is wearing a shemag (scarf) bought in a local market. On his shoulder is his personal communications handset, which is attached to his equipment by a cable. The infantryman, a member of the KMNB (Kabul Multinational Brigade), is also wearing a bulletproof vest. His CRKT Desert Cruiser can be seen in the bottom right corner of the photo. Photo: Sören Sünkler.

Photo: Sören Sünkler

CRKT DESERT CRUISER

Rugged, all-round folding knife

The Desert Cruiser is an impressive combat knife. Like many CRKT models, it is based on a custom knife design by the renowned American knife maker Jim Hammond, a member of the American Knifemaker's Guild since 1978. The Hammond Cruiser, the luxury version, was first revealed in the American magazine "Tactical Knives" in May 2003. The testers declared: "The Cruiser rocks!"

The CRKT version of the knife also "rocks", and in a number of respects. First of all is its size, for when folded the Desert Cruiser measures a considerable 13.3 centimeters. The knife therefore fits comfortably even in large hands. The well thought-out shape of the handle with deep finger choil and two shallower finger grooves ensures that the knife rests firmly and comfortably in the hand. The tough Zytel scales are radiused and contoured, the pommel is curved downward in a parrot's beak, which significantly improves grip. Beneath the Zytel scales are the proven 420J2 stainless liners (one of which locks the blade). In addition to the main lock, the Liner Lock, the Desert Cruiser is also equipped with a manually-operated LAWKS mechanism, in which the liner is blocked by pushing a lever forward.

The blade, which moves on Teflon bearings, can be opened either with the dual thumb studs (Torx screws) or with the Flipper, which projects from the open spine of the handle when the knife is closed. The Clip Point blade with long false top edge can be used in many roles, not least as a stabbing weapon. The blade is hollow-ground on both sides, however the cutting edge is biased to the left. The AUS-8 steel, which itself is relatively rustproof, is protected against corrosion by a titanium nitride coating.

The blade coating and the two Zytel handle scales are desert sand color, which results in a rather unique appearance – perfectly suited to the operations zones in Iraq and Afghanistan where the knife is often found. A sturdy metal clip (a second is included) enables the Desert Cruiser to be attached to the uniform or equipment. The clip can be mounted in four different positions on the handle. The Desert Cruiser is a heavy and rugged folding knife which is capable of standing up well against many fixed combat knives.

CRKT designer and knife maker Jim Hammond.

The Desert Cruisers clip is adjustable, the spine is open for cleaning.

SPECIFICATION

Manufacturer:	Columbia River Knife and Tool
Model:	Desert Cruiser
Blade Length:	95.0 mm
Blade Thickness:	3.5 mm
Blade Steel:	AUS-8
Overall Length:	229.0 mm
Locking:	Liner-Lock, LAWKS
Weight:	176 g
Handle:	Zytel
Manufactured in:	Taiwan
Supply Number:	none (privately procured)

Soldiers of the British armed forces were deployed in southern Iraq between 2003 and 2009 as part of "Operation Telic". The one seen here is a member of the 2nd Battalion "The Poachers", The Royal Anglian Regiment. Between May and November 2008, "The Poachers" provided personnel for Military Transition Teams during "Operation Telic 12". The role of the MTTs was to train Iraqi troops and advise and support them during operations. In addition to an SA80 bayonet, this MTT member is carrying a CRKT K.I.S.S. within easy reach on his Osprey body armor.

CRKT K.I.S.S.

Simple, small and effective

The reverse side of the KISS is equipped with a sturdy clip.

CRKT designer and knife maker Ed Halligan.

The British soldier in the photo on the left is carrying an SA80 bayonet stuck in the MOLLE loops of his combat vest plus a KISS series knife made by the American company CRKT. It is the first KISS model, which Columbia River Knife and Tool introduced in 1998. The Tanto blade with bead-blast finish was made of Japanese AUS-6 steel (since then the blade has been made of 420J2 steel). The design is based on a custom knife designed by the former aircraft mechanic and knife maker Ed Halligan. Now more than seventy years old and still active, he was one of the first to whom CRKT gave a license for quantity production. The partnership continues to this day and regularly bears fruit in the shape of numerous awards received by CRKT knives.

KISS stands for "Keep It Super Simple", and in fact the knife's simplicity is genius. Essentially it consists of just three parts: the blade, the clip and a metal frame which is both handle and lock. The Tanto blade, which can also be obtained with serrated edge grind, is chisel-point ground on the front side and flat on the backside. This chisel-point grind is in fact on the wrong side for a right-hander, but it is necessary in order to allow the blade to lie close enough to the frame when folded to prevent injuries.

The blade is 59 millimeters long, the maximum blade thickness is three millimeters. The thumb stud on the left makes it possible for a right-hander to open the blade with one hand. The blade is locked using the Liner Lock principle, in which a part of the handle under tension clicks into place behind the base of the blade in the open position. Because of its all-steel design, this small knife weighs about 65 grams and is exceptionally sturdy.

To be sure, the KISS is not the prototype of a military knife – its possible uses are limited by the short, chisel-point Tanto blade, and its 89-mm-long handle is too short. The KISS is, however, thoroughly usable as a utility knife.

The chisel-point grind allows the blade to sit close to the handle.

SPECIFICATION

Manufacturer:	Columbia River Knife & Tool
Model:	K.I.S.S. Tanto
Blade Length:	59.0 mm
Blade Thickness:	3.0 mm
Blade Steel:	AUS-6 oder 420J2
Overall Length:	147.0 mm
Locking:	Frame-Lock
Weight:	65 g
Handle:	Steel
Manufactured in:	Taiwan
Supply Number:	none (privately procured)

24 December 2007, somewhere on the Syria-Iraq border: in a forward operations base in the middle of the desert, a Marine uses his CRKT M16-13 to open a box of 24-hour rations. Between August 2007 and March 2008, the U.S. Marine Corps' 3rd Light Armored Reconnaissance Battalion operated in the border region. The battalion's mission was to prevent terrorists and resistance groups from smuggling weapons and explosives into Iraq from Syria.

CRKT M16-13 MILITARY

All-round Spear Pointer with Memory Function

Rations for the troops in the desert: beef stew or chicken and pasta? Whatever it is, it must be something of this nature which the soldier on the left is removing from a box of rations with the aid of his CRKT M16 – 13M. As to the palatability of the "Sun Meadow" rations, opinions vary widely. As to the CRKT M16–13 Military, however, the verdict is unanimous: direct hit!

The M16 series – along with the similar M4 and M21 models – was developed by Harold "Kit" Carson, a former army Sergeant Major. The man knew from experience what he was doing. Unlike the normal M16–13 models, the M version comes with an olive-green anodized aluminum handle with black Teflon-coated 420J2 steel liners. Also noteworthy is the 1* logo on the military models, which has its origins in a remark by Gary Paul Johnston which essentially meant "one ass to risk". And if one has only one behind, then one should do everything possible to protect it.

The M16-13 is an effective combat knife. It has a rugged design, has an effective locking mechanism, and is so easily attached to clothing or equipment that it can always be carried. The Spear Point blade is up to 2.5 millimeters thick and has a black titanium nitride coating, which minimizes reflection and protects the blade against rust. It is hardened to 58 to 59 Rockwells, which ensures a high degree of hardness. Only the main edge is highly sharpened, and from there to the grip the blade has neat serrations. On the top of the blade there is an extended false edge.

The removable but not repositionable clip (for tip-down carrying style) is mounted beneath the blade axis and is similarly coated. When fixed to the pants pocket, a good part of the handle projects, making it possible to draw the knife easily. Regardless of whether one uses the thumb studs (which also function as blade stops) or the Flipper – thanks to the precision manufacturing and Teflon bearings, the blade opens quickly and smoothly. The M16-13 is equipped with a Liner Lock and the LAWKS system to prevent inadvertent closing. In fact, there is nothing that can go wrong. The M16-13 is a light and rugged folding knife for normal tasks (such as opening cartons) and emergencies.

One of the identifying features of the M16 series is the Flipper, which is both an opening aid and hand protector.

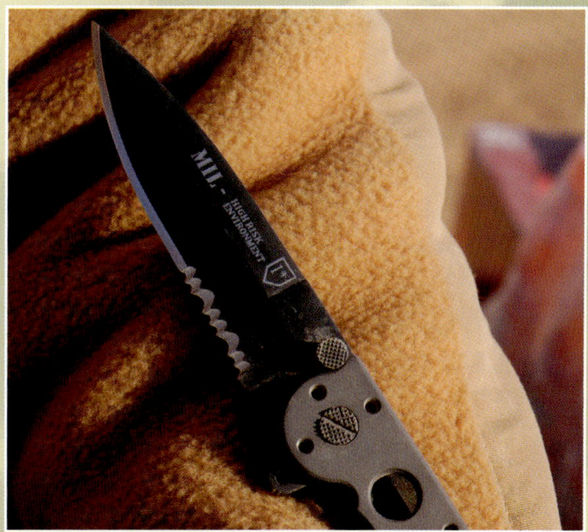

Note the chisel-point edge and the aggressive serrations.

SPECIFICATION

Manufacturer:	Columbia River Knife & Tool
Model:	M16-13M
Blade Length:	89.0 mm
Blade Thickness:	2.5 mm
Blade Steel:	AUS-8
Overall Length:	210.0 mm
Locking:	Liner Locks, LAWKS
Weight:	90 g
Handle:	6061-T6 aluminum
Manufactured in:	Taiwan
Supply Number:	none (privately procured)

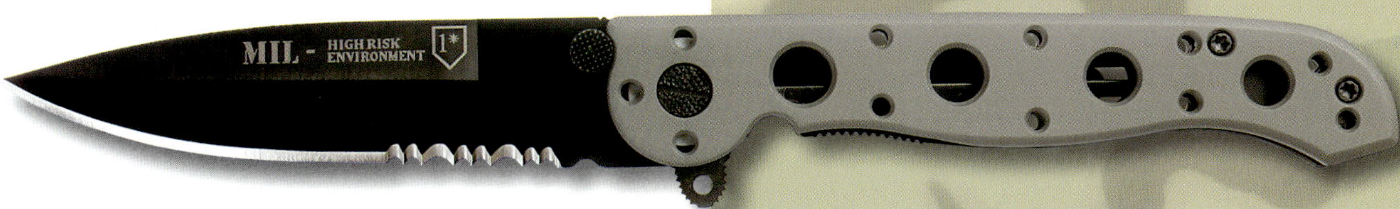

Baghdad, Iraq, April 2009. Armed with an M249 SAW 5.56-mm light machine-gun, a soldier of C Company, 2nd Battalion, 505th Parachute Infantry Regiment guards the entrance to an apartment house in the eastern part of the city. On his Interceptor Body Armor combat vest he is carrying an M16-14D by CRKT. His fellow soldiers had just searched the building for hidden weapons. The 2nd Battalion, 505th Parachute Infantry Regiment is part of the 3rd Brigade Combat Team "H Minus" of the 82nd Airborne Division "All American".

CRKT M16-14D DESERT BIG DOG

The Big Desert Dog.

CRKT designer and knife maker Harold "Kit" Carson.

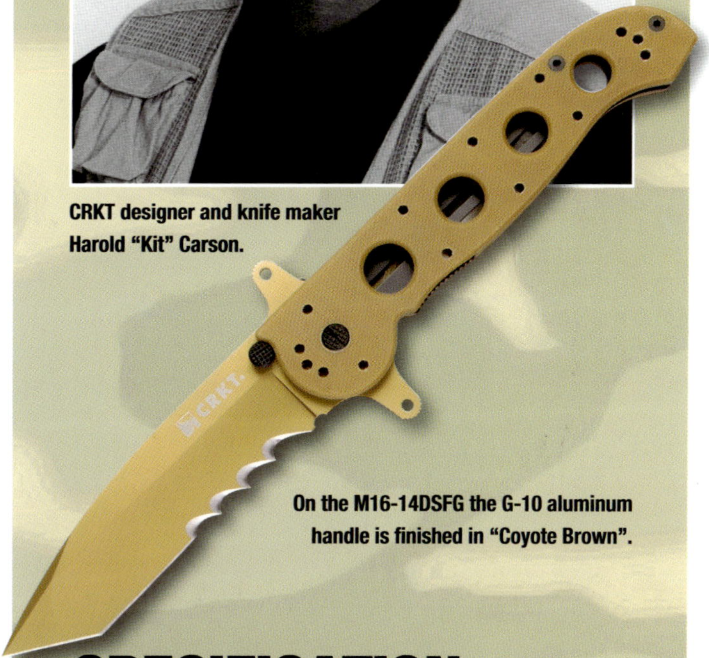

On the M16-14DSFG the G-10 aluminum handle is finished in "Coyote Brown".

The desert is one of the most difficult and dangerous of environments – to survive there one must adapt to the conditions and carry the necessary equipment. Kit Carson and Columbia River Knife and Tool developed the rugged, heavy M16-14D combat knife for the combat forces serving in the Middle East and Afghanistan, to ease the daily life of the soldiers and provide them with a knife they can rely on in emergency situations.

Folded, the knife measures almost 14 centimeters – sufficiently roomy even for large hands in gloves. The contoured grip and the flipper, which also serves as a hand protector when the blade is open, enhance the grip. This is necessary too, because the aluminum grip, while very robust (even in the skeletonized version), is also rather slippery.

The liners of rust-proof steel beneath the handle scales provide additional stability. One of the liners arrests the open blade. The AutoLAWKS mechanism, in which a catch under pressure automatically blocks the Locking Liner as soon as the knife is open, also secures the Tanto blade. The M16 series knives are thus almost as secure as fixed knifes.

The color of the blade and handle matches the prevailing color in the desert regions: the handle is hard anodized "Desert Tan", while the blade has a corrosion-resistant "Desert Tan" titanium nitride coating. Nothing shines, not even in sunshine. The Tanto blade is up to 3.5 millimeters thick, which together with the 58 to 59 HRC hardened AUS-8 steel results in increased durability. Even the pointed tip has no tendency to break off when subjected to bending loads. A double-sided hollow-ground tip gives way to a chisel-point grind edge with serrated section.

The four-position metal clip ensures a snug fit on the pants, vest or backpack. The M16-14D is east to use, thanks to the thumb studs (with checkered surfaces), the Flipper and the Teflon bearings. Because of its open-frame design, the knife does not have to be disassembled for cleaning. The CRKT M16-14D makes a huge impression – not least because of the 150 grams the bid desert dog brings to the scales.

SPECIFICATION

Manufacturer:	Columbia River Knife & Tool
Model:	M16-14D
Blade Length:	98.0 mm
Blade Thickness:	3.5 mm
Blade Steel:	AUS-8
Overall Length:	235.0 mm
Locking:	Liner-Lock, Auto LAWKS
Weight:	150 g
Handle:	6061-T6 aluminum
Manufactured in:	Taiwan
Supply Number:	none (privately procured)

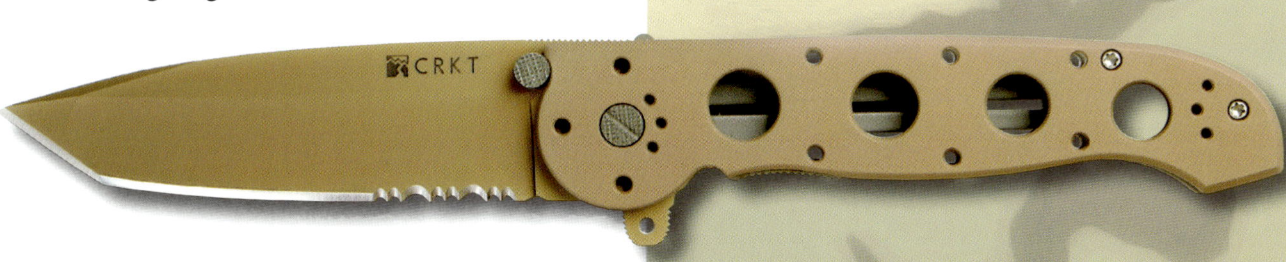

January 2008: in preparation for a mission in Baqubah, a GI enters data into a radio set. His M16-14T by CRKT lies on the ground beside him. The soldier is a member of the 4th Battalion, 9th Infantry Regiment of the 4th Stryker Brigade Combat Team of the 2nd Infantry Division. He wears the new Army Combat Uniform (ACU) with digital camouflage pattern, recently introduced by the army.

CRKT M16-14T

The High-Tech Variant of the M16-14 with Titanium Frame.

The knife lying on the ground in the photo on the left is the most versatile and expensive version of the M16-14 series, the T model with titanium frame. The knife still retains the grip of the original Custom model by knife maker "Kit" Carson.

The handle frame of the M16-14T is made of 6A14V titanium, which is more durable than aluminum and lighter than steel, resulting in an ideal strength to weight ratio. Four lightening holes in each side of the frame further reduce weight. The ceramic bead-blast finish results in a slightly rough outer surface, affording a firm grip. On the back side of the handle there is a steel clip, which allows the M16-14T to be secured tip-down on clothing or equipment. While the clip is removable, it is not adjustable. The mount beneath the blade axis leaves a relatively large portion of the knife sticking out of the pants pocket, allowing the user to grasp it quickly and securely.

In the bottom half of the frame there is a 420J2 steel liner, which locks the opened blade. Unlike the Carson Custom model, the M16-14T retains the AutoLAWKS system. When the blade is opened, a spring-operated bolt blocks the liner, making it almost impossible for the blade to be closed inadvertently.

The blade is made of Japanese AUS-8 steel, with excellent edge retention and corrosion resistance. At the same time, the blade can be sharpened with relatively simple aids. The blade has a fine bead-blast finish. The last third of the blade before the handle has an aggressive serrated edge, capable of cutting fibrous materials, heavy rope, plastic and even thin metal. Almost ten centimeters long, the Tanto blade is 3.5 millimeters thick at the spine, and the double-side hollow grind commences at the tip of the somewhat low bevel edge. The rugged point is typical of Tanto blades.

The knife can be opened with one hand using either the thumb studs or the so-called Carson Flipper. With the blade open, the Carson Flipper also provides reliable hand protection. The blade and handle are rugged, and the blade steel, hardened to 58 to 59 Rockwells, leaves nothing to be desired.

The AutoLAWKS mechanism, which locks the blade open and prevents inadvertent closing.

View from below of the blade locking mechanism of the M16-14T.

SPECIFICATION

Manufacturer:	Columbia River Knife & Tool
Model:	M16-14T
Blade Length:	98.0 mm
Blade Thickness:	3.5 mm
Blade Steel:	AUS-8
Overall Length:	235.0 mm
Locking:	Liner-Lock, Auto-LAWKS
Weight:	162 g
Handle:	Titanium
Handle:	Taiwan
Supply Number:	none (privately procured)

The mission of the UK EOD Group's "Search Troop" was to detect improvised explosive devices and mines, so that they could be disarmed by EOD teams. The UK EOD Group was a joint force, with personnel from the British army, navy and air force. This British "Search Troop" member (photographed in Basra in the summer of 2008) has attached his Gerber 06 Automatic on his combat vest within easy reach.

GERBER 06 AUTOMATIC

No Knife for Civilian Use

The 06 Automatic by the Gerber Company of Oregon has been a tremendous success. The materials used in its construction are first class and the design is well thought-out and specially suited to meet the demands of the police and military. The knife can only be sold to members of these organizations in the United States.

The 06, here secured to the vest loops, is assembled entirely with Torx screws. The machine-turned handle is made of hard-anodized (Mil-Spec Type III) 6061-T6 aluminum. Three finger grooves ensure that the knife sits firmly in the hand. At the blade end, the handle provides protection for the hand. Oblique grooves and riffling on the sides of the handle significantly improve grip firmness, whether the blade is being used or the 06 is being employed as a striking aid. The 06 was designed specifically for this: there is a strike point in the pommel which, like a Kubotan, concentrates striking energy onto a small area. The pommel also has a hole for a lanyard.

The knife is delivered with a nylon case with Velcro fastener, which fits all available belts. The 06 Automatic is also equipped with a metal clip, which enables the knife to be attached to clothing or equipment tip-up. The sturdy clip is attached to the handle by three Torx screws and can be transferred to the other side of the grip.

The blade opens automatically at the push of a button. The opening button has intentionally been made very large, so that the knife can be operated even when the user is wearing gloves or under stress. The knife is provided with a slide safety to prevent inadvertent opening or closing. With it the button can be blocked with the blade in the open or closed position. S30-V black oxide coated stainless steel is used for the drop-point blade (there is also a Tanto version, also with partial serrated edge). With a carbon content of 1.45 percent and chromium and vanadium components (14% and 4% respectively), this steel is very wear-resistant, has excellent edge retention and is also relatively strong. Blade thickness is a considerable 4.7 millimeters.

The alternative Gerber 06FAST comes with a handle made of G-10 composite material.

The strike point on the pommel can be used as glass breaker.

SPECIFICATION

Manufacturer:	Gerber Legendary Blades
Model:	06 Automatic Drop Point
Blade Length:	96.0 mm
Blade Thickness:	4.7 mm
Blade Steel:	CPM-S30V
Overall Length:	218.0 mm
Locking:	Button-Lock
Weight:	—
Handle:	6061-T6 aluminum
Alternatives:	Tanto, G-10-Model
Manufactured in:	USA
Supply Number:	NSN 1095-01-446-4348

A machine-gunner of A Company, 1st Battalion, 38th Infantry Regiment poses for the camera with his Gerber A-F Combat Folder in Baqubah in January 2008. He is armed with an M249 SAW 5.56-mm light machine-gun. On the weapon is an EO-Tech reflector sight. For a considerable period of time, Baqubah was the scene of frequent engagements between US troops and terrorists and regional resistance groups.

GERBER APPLEGATE FAIRBAIRN COMBAT FOLDER

Tactical Combat Dagger with Roots in the Second World War.

The dagger blade of the Gerber folding knife that resulted from a collaboration between close-quarters combat expert Colonel Rex Applegate and knife makers William Harsey and Butch Valotton is a full 11.4 centimeters long. The matt-finished blade bears the names of the former OSS colonel and his British SOE instructor comrade William E. Fairbairn, one of the most important close-quarters combat experts of the Second World War. Together they developed the fixed variant of the Applegate-Fairbairn combat knife. The folding version first appeared in 1996 and immediately achieved impressive sales figures.

The blade is made of corrosion-resistant 440C steel, which is hardened to 59 Rockwells. This steel has good edge retention and corrosion resistance characteristics. The first third of the cutting edge is serrated. The spear-style blade is ground on just one side, but a practiced hand can rework it into a full-fledged dagger. With a maximum blade thickness of 3.8 millimeters, the A-F Combat Folder is quite rugged for a dagger, at least for the first quarter, for from there to the tip the blade becomes progressively thinner. The dual thumb stud openers and the belt clip, which can be mounted on either side, make the knife ideal for right- and left-handers.

The handle measures 14.3 centimeters, making the knife almost 26 centimeters long in the open position. The center of gravity is about one centimeter behind the axis in the direction of the pommel. Inside the fiberglass-reinforced Zytel nylon handle scales are steel liners, one of which locks the blade. The liner lock can also be blocked by a slide safety. The surface of the handle is textured and the lengthwise riffles improve the grip considerably. At the front of the handle are two riffled grooves for the index finger and thumb, and the rear part of the blade spine is also provided with anti-slip riffles. A lanyard can be attached to the pommel.

The A-F Combat Folder was designed primarily as a fighting knife, however it is also suitable for normal cutting jobs. Because of its geometry, the blade might be too liable to break under heavy field use.

Col. Rex Applegate and Dietmar Pohl during their first meeting in Dallas in 1996.

The A-F Combat Folder is also available with a Foliage Green handle.

SPECIFICATION

Manufacturer:	Gerber Legendary Blades
Model:	A-F Combat Folder
Blade Length:	114.0 mm
Blade Thickness:	3.8 mm
Blade Steel:	440C
Overall Length:	257.0 mm
Locking:	Liner-Lock
Weight:	184 g
Handle:	Zytel
Options:	Covert, Mini-Version
Manufactured in:	USA
Supply Number:	none (privately procured)

The Interceptor Body Armor (BA) combat vest with Improved Outer Tactical Vest (IOTV) is the U.S. Army's standard combat vest today. Many American soldiers serving in Iraq and Afghanistan in recent years owe their lives to this vest. With front and back plates (ESAPI) and side plates (ESBI), groin protector, upper arm and neck protector, the combat vest weighs about 15 kg. The vest depicted here belongs to a specialist of the 4th Stryker Brigade Combat Team of the 2nd Infantry Division. On the front part of the vest he carries into action shotgun shells, a carbine hook and a Gerber E-Z Out Skeleton.

GERBER E-Z OUT SKELETON

Everything Easy – A Simple Knife for Simple Tasks

"E-Z to access, E-Z to open and E-Z to use!" Thus reads the motto of the E-Z Out Skeleton series by Gerber. And in fact, thanks to the rugged pocket clip of corrosion-proof steel which is attached to the reverse side of the handle by the axis pin, the knife can easily be affixed to clothing or items of equipment – making it easy to reach at all times. The clip is not movable, however. If the knife is clipped inside the pants pocket, a good centimeter of the handle projects beyond the edge of the pants. Bad if one doesn't wish to attract so much attention. Good if one wants to draw the knife quickly and safely.

Because of its perforations, the blade of the E-Z Out Skeleton can easily be opened one-handed, by both right- and left-handers. With a little skill, the blade can also be closed one-handed by pressing down on the centrally-located Back-Lock release. Thanks to the large handle with molded Kraton rubber scales, the knife fits comfortably into the hand. Moisture has little impact on grip. This handle composition has proved itself for decades on Gerber's Gator series. The knife is also relatively light, thanks to the absence of a stabilizing steel liner and the large perforations in the blade. The drop-point blade of the E-Z Out Skeleton is hollow ground, and the edge has sufficient belly for pulling cuts.

The E-Z Out Skeleton is basically a very simple knife which is adequate for normal cutting tasks. For heavier work, the relatively narrow and perforated blade reaches its limits. The edge retention qualities of the steel used by Gerber (400 Series) is also not particularly high. On the other hand, the Back-Lock knife is relatively light and the comfortable grip is hard to beat. The Gerber E-Z Out Skeleton is certainly not the prototype of a military knife on which one can rely in extreme situations. But, as one sees, even the simplest folding knives find their way into war zones. The E-Z Out Skeleton is entirely suitable for cutting a rope, preparing food or cutting oneself free. It is especially useful if one is already equipped with a bayonet or fixed fighting knife with which to carry out heavier tasks.

For many soldiers, one knife is not enough. In addition to an E-Z Out Skeleton, this one is carrying an A-F Combat Folder and a Multi-Tool.

SPECIFICATION

Manufacturer:	Gerber Legendary Blades
Model:	E-Z Out Skeleton
Blade Length:	97.0 mm
Blade Thickness:	3.5 mm
Blade Steel:	AUS-8
Overall Length:	232.0 mm
Locking:	Back-Lock
Weight:	216 g
Handle:	Zytel
Manufactured in:	Taiwan
Supply Number:	none (privately procured)

The crews of the LAV-25 wheeled armored vehicles of the USMC's 3rd Light Armored Reconnaissance Battalion include four Ground Scouts. They are armed with M16A4 5.56-mm assault rifles, M203 40-mm grenade launchers and M249 SAW 5.56-mm light machine-guns. This Ground Scout is carrying a Gerber Paraframe II folding knife on his Modular Tactical Vest (MTV).

GERBER PARAFRAME

Light Frame-Lock Knife of Modern Design

The folding knife the soldier is carrying in the MOLLE loops of his ballistic vest is a Gerber Paraframe I. The American company Gerber Legendary Blades was founded in Portland, Oregon in 1939 (since then Gerber has been bought out by Finnish tool specialist Fiskar) and during its existence has produced several legendary knives: the Vietnam Combat Knife Mark II, the Gator Series models and the Applegate-Fairbairn Covert. Though it has not achieved legendary status, the Paraframe is a reasonable and good-functioning one-handed knife..

The open frame design of the rustproof steel handle contributes to the knife's low weight. The Clip-Point blade is locked by a Frame-Lock mechanism. This locking mechanism, in which part of the frame slides behind the base of the extended blade, was developed by Chris Reeve, known mainly for his one-piece survival knives. The two halves of the handle are held together with Torx screws, allowing the knife to be disassembled for cleaning. An unspecified "surgical steel", probably 420HC or 440A, is used for the hollow-ground blade. The Clip-Point shape, in which the spine of the blade has a concave edge to the tip, has advantages and disadvantages. Having the tip at the height of the center of the blade allows for controlled work, and one knows intuitively where one is stabbing. The thinness of the blade in the tip area also makes fine cuts possible. The fineness of the blade is also a disadvantage, however: it breaks or bends easily, depending on the blade material and hardness.

The Paraframe I is available with a fine edge or half-serrated blade. Advantages of the serrated blade are its better cutting capability with fibrous material and heavy rope, and as well it remains sharp for a relatively long time. When the time comes to sharpen it, however, it is difficult. There is still no 100% effective sharpening system. Twin thumb studs make it possible to open the knife one-handed, with the right or left hand. The long clip, on the other hand, is not adjustable and is designed for the carriage methods used by right-handers.

For military users the Paraframe model is also available with black coating.

The highly polished surfaces of the Paraframe produce annoying reflections.

SPECIFICATION

Manufacturer:	Gerber Legendary Blades
Model:	Paraframe I
Blade Length:	76.0 mm
Blade Thickness:	2.7 mm
Blade Steel:	400
Overall Length:	180.0 mm
Locking:	Frame-Lock
Weight:	Not known
Handle:	Steel
Options:	Mini Paraframe, Paraframe II (blades: 57 and 89 mm, respectively)
Manufactured in:	Taiwan
Supply Number:	none (privately procured)

Specialist Shaw is a member of the Military Police Platoon of the Special Troop Battalion of the 3rd Brigade Combat Team of the 92nd Airborne Division "All American". In March 2009 the unit was deployed in the eastern part of Baghdad. While on standby the soldiers spent their time close to their vehicles. Sitting on the steps of the rear ladder of a Caiman MRAP, Specialist Shaw is using the blade tip of her Gerber Torch II Tanto to clean her fingernails.

GERBER TORCH II TANTO

Folding Tanto for Many Uses

Clichés can be so apt: here a female American soldier is using her big, black Tactical Folder, not to pry open an ammunition box, but to clean her fingernails. The Torch II Tanto is obviously a true all-rounder – not just a functional folding knife for tactical use, but also suitable for hand care. The knife was designed by American knife maker Rick Hinderer, who is a blacksmith by trade and also a lifeguard and volunteer fireman in his hometown in Ohio. Hinderer had previously worked with Gerber, and the Hinderer Rescue, in particular, was a great success. The design of the Torch is based on a custom knife model by Rick Hinderer, of which only a few examples were made.

Made in Taiwan, the production knife has a handle of black-coated steel with inserts of G-10 composite material. Closed, the knife measures 11.8 centimeters. Notches on the top and bottom of the handle end further improve the grip. An oval hole in the handle makes it possible to attach a lanyard. There is a metal clip on the reverse side of the handle for attaching the knife to clothing. The clip is removable, but not adjustable.

Additional hand protection with the blade open is provided by the finger push. The finger push allows the blade of the Torch to be snapped open. With the blade folded, the finger push extends from the bottom side of the handle level with the blade axis. Alternatively, the knife can also be opened with the dual thumb studs. Locking of the nine-centimeter blade is achieved with a sturdy Frame-Lock, which even after decades is still one of the safest locking mechanisms extant.

The blade – in typical Gerber jargon – is made of rust-proof surgical steel (probably 440A) and has a Tanto point. The blade is coated black to minimize reflection. Before the serrated section, the blade has a concave area which can be used as a finger groove if one wishes to shorten the grip. The edge is hollow-ground, and there is a false edge ground on the spine of the blade.

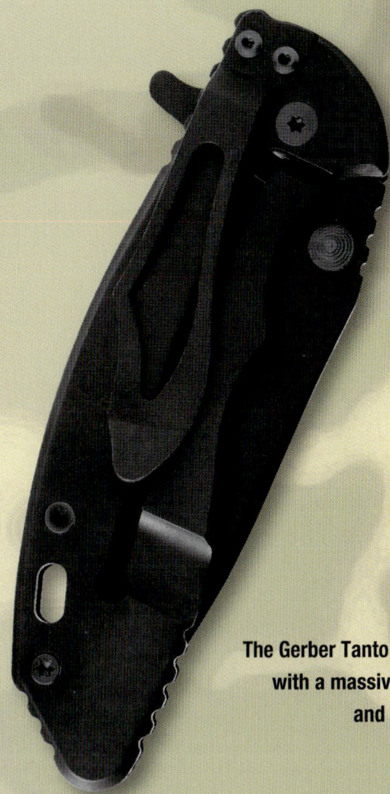

The Gerber Tanto II is equipped with a massive Frame-Lock and a trouser clip.

SPECIFICATION

Manufacturer:	Gerber Legendary Blades
Model:	Torch II Tanto
Blade Length:	90.0 mm
Blade Thickness:	3.5 mm
Blade Steel:	440A
Overall Length:	208.0 mm
Locking:	Frame-Lock
Weight:	130 g
Handle:	Steel/G-10
Manufactured in:	Taiwan
Supply Number:	none (privately procured)

This LAV-25 commander of the USMC's 3rd Light Armored Reconnaissance Battalion relies on a Ka-Bar Mule in action. The crew of a LAV-25 wheeled armored vehicle consists of the commander, the gunner and the driver. As many as six ground Scouts can be carried in the vehicle's aft fighting compartment.

KA-BAR MULE

Tough Workhorse for All Operational Purposes

Mule is a fitting name for this folding knife made by the American manufacturer Ka-Bar. Like its animal namesake, it is tough, robust and simple. The knife's rugged design is also reflected in its weight of about 200 grams. Its two steel liners, each 1.5 millimeters thick, the powerful back-lock spring and the almost ten-centimeter-long and 3.5-millimeter-thick blade all contribute to the knife's impressive weight. The hollow-ground blade of AUS-8 steel is available in Clip-Point or Tanto shape. The black Teflon coating increases corrosion resistance and minimizes reflections. The soldier depicted here obviously uses his knife frequently and for hard work, as indicated by the scratched coating.

The knife can be operated one-handed, even when wearing heavy gloves. The double thumb studs and adjustable clip (tip-up) make the Mule suitable for left-handers. The powerful back spring produces considerable resistance when opening the blade. Once opened, however, there is absolutely no play in the blade, which is as it should be. The locking hammer is released by means of a Boye Dent, designed by developer David Boye. The recess performs two functions: on one hand it prevents the inadvertent depression of the back spring during a firm grip, on the other it provides the thumb with a sure grip during unlocking.

The handle is wider toward the butt, which effectively prevents the knife from sliding forward out of the hand. Towards the blade, grip is improved by an aggressively riffled finger plate and a deep finger groove. When clipped, only a little of the handle projects from the pants pocket. Because of its weight, however, the knife is better carried in a belt sheath.

Two Zytel scales are screwed to the steel liners. The knife can be disassembled into its basic components for cleaning. The rounded and three-dimensional handle (there are recesses in the sides at the front of the handle) rests comfortably and securely in the hand. Slip-proof Kraton inserts further improve the grip. The Mule's blade has a usable length of 9.7 centimeters, sufficient for almost every type of use. It is a true workhorse for military users.

The Mule is also available with a Tanto blade.

The Ka-Bar Mule sheath holds the heavy folding knife securely.

SPECIFICATION

Manufacturer:	Ka-Bar Knives, Inc.
Model:	Mule, Clip Point
Blade Length:	97.0 mm
Blade Thickness:	3.5 mm
Blade Steel:	AUS-8
Overall Length:	232,0 mm
Locking:	Back-Lock
Weight:	216 g
Handle:	Steel/Zytel/Kraton
Alternatives:	Tanto
Manufactured in:	Taiwan
Supply Number:	none (privately procured)

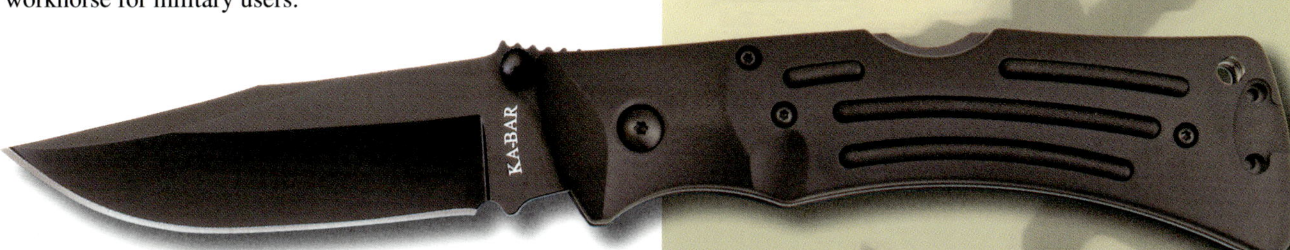

Lance Corporal Smalley is a Ground Scout with the USMC's 3rd Light Armored Reconnaissance Battalion. His battalion's area of operations is western Iraq. On the shoulder straps of his combat vest he is carrying a tourniquet strap (left) and a Ka-Bar Desert Mule.

KA-BAR MULE DESERT & FOLIAGE

Folding Knife Colored to Match Dusty and Marshy Areas

These models of the Mule series feature a Zytel handle finished in the colors desert-sand or swamp-green. Four lengthwise Kraton inserts improve the grip. The camo version of the Mule sits securely in the hand, thanks in part to the handle's two finger grooves, its widened butt and riffled thumb rest. The handle measures 13.3 centimeters – the benefits of this added length become apparent when working with gloves. At 216 grams, the knife is not exactly a lightweight. The two 1.5-millimeter-thick steel liners, to which the Zytel scales are screwed, a powerful Back-Lock spring and the 9.7-centimeter-long and 3.5-millimeter-thick blade all contribute to the knife's weight. The knife's mass exudes reliability, however.

Made of AUS-8 steel, the blade is available in Clip-Point or Tanto style. The blade of both versions is hollow ground, and the knives come from the factory with a grind angle of 15 degrees (30 degrees total). The black, wear-resistant Teflon coating improves resistance to corrosion. In addition to other alloy components, the Japanese AUS-8 steel contains 0.7 percent carbon and about 14 percent chromium. This type of steel is hardened to 57 to 59 Rockwells, ensuring that the blade has sufficient bite but does not break too easily.

The large dual thumb studs and adjustable clip make it easy for left-handers to use the Mule models. One does need quite a lot of strength to open the knife one-handed, however, as the spine spring produces considerable resistance during opening. This is typical of all Back-Lock knives though. The clip is positioned and shaped so that virtually none of the knife is visible when it is carried clipped. Because of its weight, however, it is recommended that the Mule be carried in the sheath that comes with the knife. The latter matches the knife, being either sand or foliage color. The bi-positionable sheath can be worn vertically or horizontally on the belt. The long closing tab is secured by a Velcro fastener and a snap.

The Desert Mule's mount sheath can be worn horizontally or vertically.

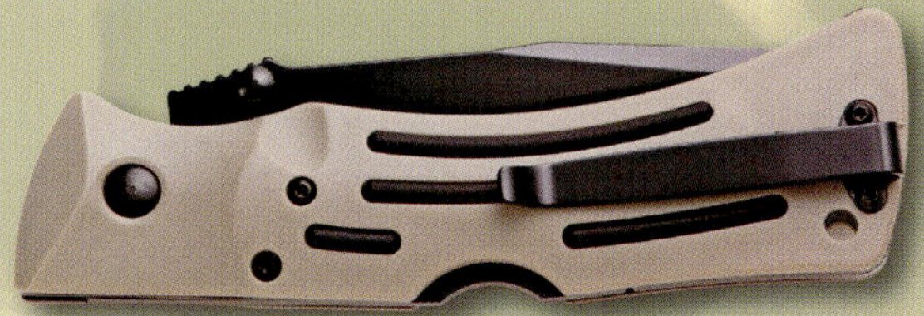

The clip of the Mule series is adjustable.

SPECIFICATION

Manufacturer:	Ka-Bar Knives, Inc.
Model:	Mule, Tanto, Desert & Foliage
Blade Length:	97.0 mm
Blade Thickness:	3.5 mm
Blade Steel:	AUS-8
Overall Length:	232.0 mm
Locking:	Back-Lock
Weight:	216 g
Handle:	Steel/Zytel/Kraton
Options:	Drop Point blade, black handle
Manufactured in:	Taiwan
Supply Number:	none (privately procured)

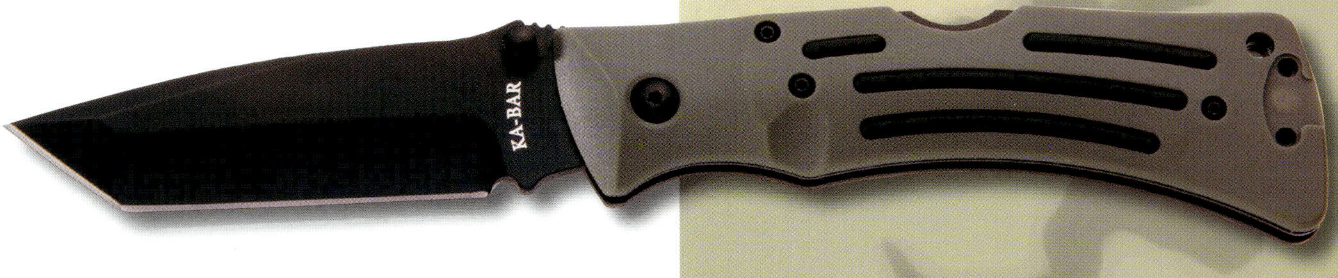

This United States Marine Corps Sergeant is the driver of an MTVR medium truck (Medium Tactical Vehicle Replacement) of the 2nd Regimental Combat Team (RCT 2). Between August 2007 and March 2008 the brigade-strength unit was responsible for security in Al Anbar province in western Iraq. RCT 2 often patrolled the Iraqi desert for weeks before returning to base. They were supplied by so-called Combat Logistics Patrols. The sergeant's MTVR was part of one such patrol. He posed for the camera in December 2007 in the Iraqi desert, holding his M-Tech one-hand folding knife. His knife has the black blade.

M-TECH RHINO FOLDING FIGHTER

Heavy Combat Knife at a Low Price

The folding knife with nylon sheath the sergeant in the photo is wearing on his left shoulder confirms that many soldiers see no need to provide themselves with the most expensive knives – or are unable to do so for financial reasons. In the latter case they must look for reasonably-priced alternatives that are nonetheless reliable.

This knife is the Rhino Folding Fighter produced by the M-Tech company (which belongs to Master Cutlery, the largest importer of swords and knives in the USA). The knife is produced exclusively in China, which makes its price very reasonable – most M-Tech models cost less than 20 Euros (25 USD). The "Rhino Folding" folding knife is one of the heaviest and rugged made by the company. The 10.8-cm blade is made of 440A steel: not especially harder or capable of holding an edge, but it is a good-natured steel which is easily sharpened and stands up to rough treatment.

Maximum blade thickness is 3.5 millimeters. Thanks to the low-set bevel, this strength is retained over much of the blade, making it very rugged. Twin thumb studs allow the blade to be opened single-handed by right- and left-handed users. A Liner-Lock locks the blade in the opened position.

The knife's heavy weight of 340 grams (!) is mainly due to the design of the handle. While the two massive steel liners with G-10 scales and the steel handle spacers are very sturdy, they are also heavy. With a length of 14.6 centimeters, the Folding Fighter is also rather on the large side. Opened, the Rhino Folding Fighter measures an impressive 25.3 centimeters – larger than many fixed knives.

The knife is assembled with Torx screws, allowing it to be completely disassembled for cleaning. A lanyard can be attached to the pommel. The heavy folding knife has no clip, but it does come with a nylon sheath with Velcro fastener, designed to be worn on the belt. With the aid of zip ties or some Paracord, however, it can easily be attached to the loops of a tactical vest.

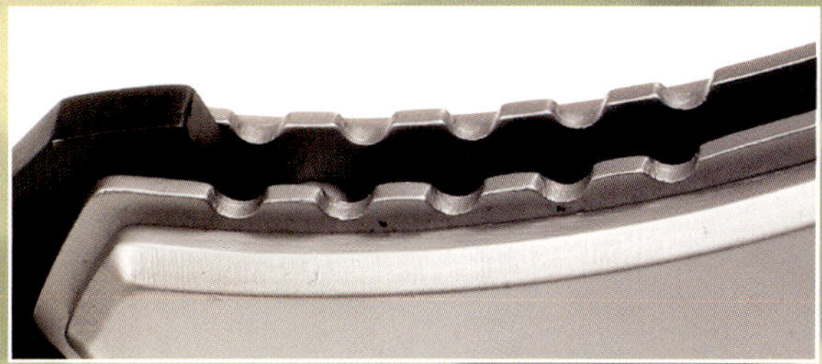

A coarsely riffled thumb rest guarantees a firm grip.

The M-Tech folding knife's Liner-Lock is very large and sturdy.

SPECIFICATION

Manufacturer:	M-Tech, Master Cutlery, Inc.
Model:	Rhino Folding Fighter
Blade Length:	108.0 mm
Blade Thickness:	3.5 mm
Blade Steel:	440A
Overall Length:	254.0 mm
Locking:	Liner-Lock
Weight:	340 g
Handle:	Steel/G-10
Manufactured in:	China
Supply Number:	none (privately procured)

Afghanistan 2009: a German ISAF soldier armed with a Heckler & Koch MP7 submachine gun and an H3 Alpha One Tactical. The knife sits within easy reach on his tactical vest in a holster specially designed for the knife by the firm 75 Tactical.

(Photo: T.M.)

POHL FORCE H3 ALPHA ONE TACTICAL

Designed for Use in the Field

Afghanistan: Pohl Force supports German ISAF forces with the H3 Alpha One Tactical. *(Photo: T.M.)*

The H3 Alpha One by Pohl Force was designed primarily for use while wearing gloves. For this reason, the handle of the Alpha One is intentionally larger than that of most folding knives. To ensure a firm grip, the handle scales, made of fiberglass-reinforced Zytel, are completely covered with molded recesses. The handle design allows for four different grips, regardless of the situation. The deep index finger groove prevents the hand from slipping onto the blade. Riffling at the front and back of the spine provide firm support for the thumb regardless of gripping position.

In selecting the blade shape, from the beginning Pohl rejected extreme Tanto or dagger shapes, which are primarily used in pure weapons. The optimal solution was found to be a spear point blade with a slight recurve. To enable the Alpha One to be used in survival situations and for light chopping tasks, the blade was made more than eleven centimeters long. The spine was made five millimeters thick to enable the blade to endure extreme loads and the cutting edge was made conical in shape. The blade is made of 440C steel, which has been used in the knife industry for more than three decades and has a good reputation with respect to edge retention, durability and resistance to corrosion.

There are many reliable locking systems, but one of the most well-proven is the so-called Back-Lock. Pohl is one of a very few manufacturers to use a safety recess on the lock. This prevents inadvertent operation of the blade lock and the recess also serves as a tactile orientation point in the dark. Even with gloves on, the user can feel the proper spot to unlock the blade. Even under stress, the user can easily close the blade. If the knife is gripped properly, the blade retracts into a safety recess without endangering the index finger.

To achieve the desired ruggedness, the Alpha-One is equipped with two liners made of two-millimeter-thick stain-

Alpha-One sheath-holster in sand color.

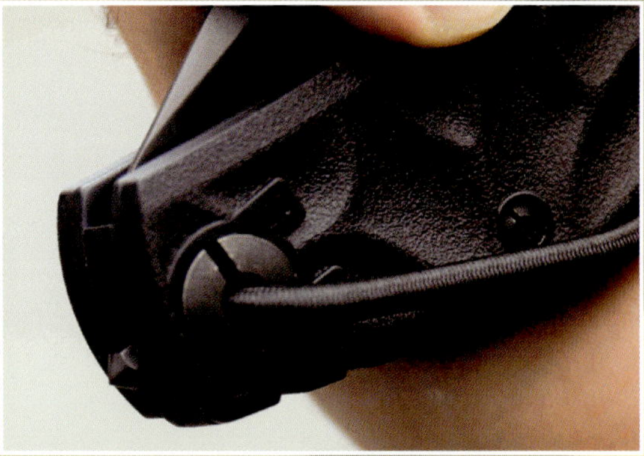

The Alpha-Two version also has a glass-breaker.

All of the screws on the Alpha One/Two can be removed with a simple screwdriver.

less steel. Cutouts in the liners keep the knife's weight within acceptable limits. Two handle pivots, with an impressive diameter of eleven millimeters, plus six slotted screws of rustproof steel ensure that all parts fit together firmly.

The two holes in the handle have another function: the Flexcord lanyard that comes with the knife can be used to secure it to the wrist in extreme situations. The cord is run through the two holes and then knotted. On both sides of the handle are sprung clips, through which the user can shove his hand. This is a practical solution if one must open the hand, for example to support oneself while climbing. The knife automatically remains in the hand.

The Alpha One's pants clip can be attached in four positions. This allows the closed knife to be carried with the blade tip-up or tip-down. The Alpha One can be used by right- and left-handers. Changing the clip can be accomplished without special tools. One needs only a coin or standard slotted screwdriver.

The improved Alpha version (Alpha-Two, from late 2009) is also equipped with a glass-breaker. This feature was added to meet the requirements of special police units.

The size of the Alpha One/Two also makes an operational sheath-holster necessary. This was also added at the behest of the special units, who requested an operational sheath-holster that could be attached to their equipment. An operational sheath-holster was subsequently developed by Firma 75. It can be operated while wearing gloves and is MOLLE-adaptable. The sheath-holster is produced in black, olive and desert colors to cover every conceivable region.

SPECIFICATION

Manufacturer:	Pohl Force GmbH
Model:	H3 Alpha One Tactical
Blade Length:	113.0 mm
Blade Thickness:	5.0 mm
Blade Steel:	440C
Overall Length:	260.0 mm
Locking:	Back-Lock
Weight:	288 g
Handle:	Zytel
Manufactured in:	Taiwan
Supply Number:	none (privately procured)

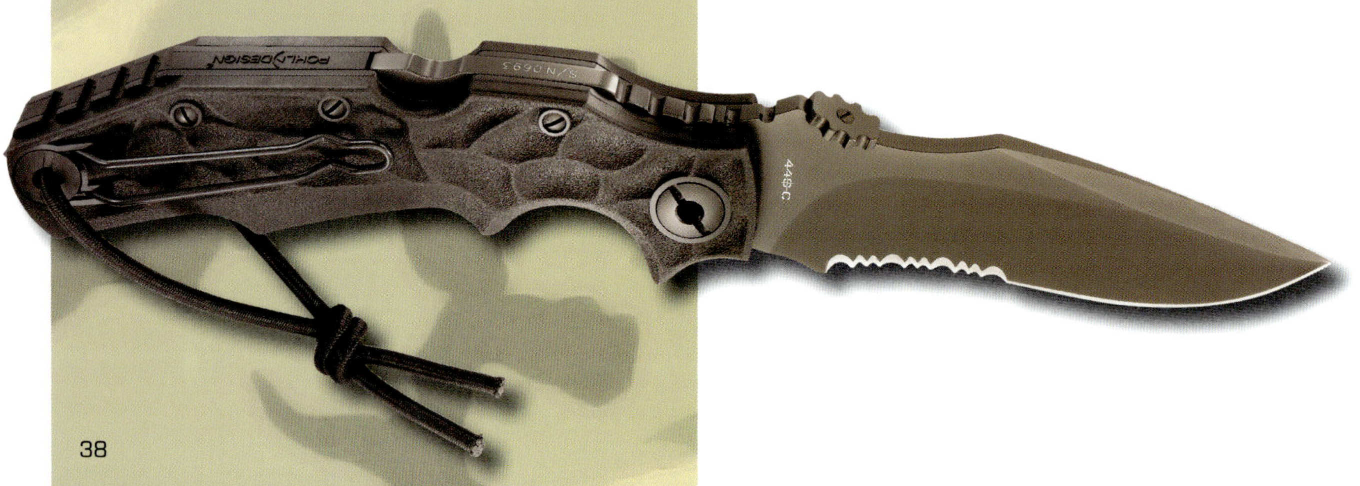

A German ISAF soldier in a camouflaged position with G22 sniper rifle. To his right is an H3 Alpha One Tactical. *(Foto: T. M.)*

Camp Pendelton 2009: a U.S. Marine shows off his H3 Alpha One Tactical which he purchased just before he was deployed to Iraq. *(Photo: Dan Erber)*

29 December 2007: U.S. Marines of the 3rd Battalion, 23rd Marine Regiment search an abandoned building on the outskirts of Haditha, Iraq. Using his M16A4 assault rifle, this marine covers the advance up the building's staircase. He has attached his Smith & Wesson Extreme Ops Automatic Knife to one of his ammunition pouches.

SMITH & WESSON EXTREME OPS AUTOMATIC KNIFE

Ready for Action at the Push of a Button

Smith & Wesson, the largest American manufacturer of firearms, quite obviously enjoys a good reputation among members of the U.S. Army, for more than a few soldiers rely on this Smith & Wesson knife, which is made under license by Taylor Brands. It is quite possible that its popularity may also be due to its price-performance ratio: the knife shown here, Model SW50BS, which the soldier has attached to the MOLLE loops of his tactical vest, can be had for about forty dollars in American on-line shops. And not surprisingly, the marine's Oakley sunglasses are also American-made.

The first third of the powder-coated drop-point blade (the Automatic series also includes Tanto blades and uncoated models) has an unsharpened false edge. This allows a fine edge on the tip area: good for fine cutting tasks and work with the tip. The model is made with half the blade plain and half serrated. The knife axis and the handle are assembled using Torx screws, and the handle scales are made of durable 6061-T6 aluminum.

Within the handle (hooked onto the hinge pin) is a powerful spiral spring, which is under tension when the knife is closed. Pressing the button withdraws the locking pin and the blade snaps out of the handle. It is locked in the open position by a Button Lock. Pressing the button again unlocks the blade, which can then be closed again under tension from the spring.

The lengthwise grooves in the handle and the finger grooves provide a reassuringly firm grip. A steel pocket clip is attached to the reverse side of the handle, and this secures the Extreme Ops Auto to clothing or equipment with the tip facing down (Tip-Down). The clip is not adjustable and is designed for right-handers.

With its relatively low price, the Extreme Ops Auto offers good value for money. As many soldiers are of the view that the military is responsible for their equipment, very few are willing to spend a lot of money on expensive knives which all too often get lost in action.

The Smith & Wesson Extreme Ops' silver-colored release button is easy to spot.

SPECIFICATION

Manufacturer:	Smith & Wesson/Taylor Brands LCC.
Model:	SW50BS
Blade Length:	83.0 mm
Blade Thickness:	3.5 mm
Blade Steel:	440C
Overall Length:	198.0 mm
Locking:	Button-Lock
Weight:	—
Handle:	Aluminum
Options:	Tanto
Manufactured in:	China
Supply Number:	none (privately procured))

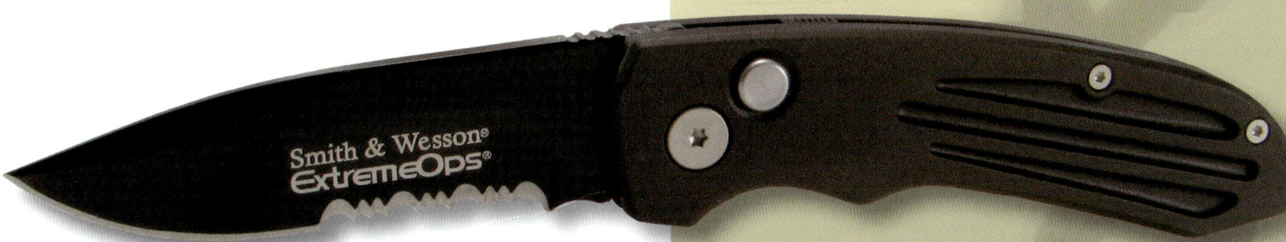

Baghdad, February 2008: in the cover of his M1126 Stryker ICV wheeled armored vehicle, an infantryman of the 1st Battalion 23rd Infantry Regiment "Tomahawks", prepares to storm a house suspected to be harboring enemy fighters. This G.I. is in the process of loading a high-explosive round into the M203 20-mm grenade launcher located under his M4A1 5.56-mm rifle. On his tactical vest he is carrying a Smith & Wesson Auto Opening S.W.A.T. 4 knife.

SMITH & WESSON AUTO OPENING SWAT 4

Fast Operational Helper for Professionals

This model from the SWAT series by Smith & Wesson has a one-piece Kraton handle insert.

In many states in the USA, the purchase and possession of automatic knives is only permitted for members of certain professional groups. These include law enforcement officers, firefighters, rescue service personnel and active members of the military. The Auto Opening SWAT 4 is only obtainable by these target groups.

The knife in the photo, visible beneath the Colt assault rifle with mounted M203 grenade launcher (next to the Cordura carrying pouch) is the automatic version of the Medium SWAT 4 by Smith & Wesson. The Auto Opening SWAT 4 fits comfortably and securely in the hand, thanks to the generous size of its aluminum handle, which has a matte chrome finish for scratch protection, and four Kraton inserts. A hole is provided in the handle for insertion of a lanyard. The narrow pocket clip covers about half the length of the handle. It has two cutouts and is attached to the handle liners by two Torx screws. The knife is assembled with Inbus screws and can be disassembled for cleaning (or replacement of the spring).

The blade is made of rustproof 440C steel. The blade spine, which has a false edge extending to the tip, curves downward (drop-point). Smith & Wesson chose a hollow grind for improved cutting qualities. The satin finish makes the knife very easy to care for while in use. Minor rust spots are easily polished away. This is a disadvantage of satin-finished blades, as rubbing away the rust spots results in unattractive shiny spots which are immediately visible. Very close to the handle there is a thumb rest with several riffles, which makes controlled cutting even safer. The single thumb stud is designed for use by right-handers. For automatic opening of the blade a hidden unlocking button beneath the forward handle insert must be pressed.

The reasonably-priced Smith & Wesson is a very popular tactical operational knife with spring-assisted opening. Although the design is no longer at the cutting edge, it is still a trusted favorite.

SPECIFICATION

Manufacturer:	Smith & Wesson/Taylor Brands LCC.
Model:	Auto Opening SWAT 4
Blade Length:	81.0 mm
Blade Thickness:	3.5 mm
Blade Steel:	440C
Overall Length:	183.0 mm
Locking:	Liner-Lock
Weight:	—
Handle:	Aluminum
Manufactured in:	China
Supply Number:	none (privately procured))

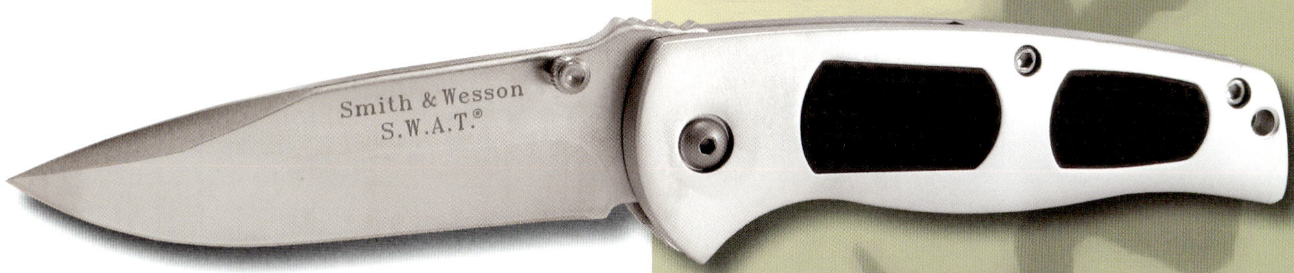

U.S. Army infantry squads are nine men strong and consist of a squad leader plus two fire teams of four men each. Each fire team consists of a team leader, a rifleman, a grenadier and a machine-gunner. This Private First Class is the machine-gunner of a fire team from B Company, 1st Battalion, 35th Armored Regiment "Iron Knights". His equipment includes an SOG Trident Tiger Stripe one-handed knife.

SOG TRIDENT TIGER STRIPE

An Extremely Quick and Lethal Tiger

Produced by the American company SOG, the Trident is considered a reliable and functional knife. The blade must first be opened slightly before the SAT spring mechanism (SOG Assisted Technology) engages and the blade snaps open. The Trident can be operated single-handed using the dual thumb studs, which also serve as blade stops. The blade is locked in the open position by means of an Arc-Lock, in which spring-assisted pins are pushed behind the base of the blade. This is a reliable and proven locking mechanism, which is also easily unlocked by right- or left-handers. There is also a slide safety to prevent inadvertent opening of the blade – only when it is in the red position is the Trident ready to "fire".

AUS-8 steel is used for the blade. With a 0.7 percent carbon content and 14 percent chrome, the steel has good edge retention and corrosion resistance qualities. The blade measures 9.5 centimeters from tip to handle, and the available sheath is almost exactly the same length. Maximum blade thickness is 3.2 millimeters. The long thumb rest has moderate riffeling. From its apex, the spine drops in a concave curve to the pointed and relatively thin tip. The durable titanium-nitride coating, which is treated in places to produce the tiger stripe effect, protects the blade against rusting. The Trident Tiger Stripe is only available with partial serrated edge.

The handle, which is just over twelve centimeters long, is made of Zytel, a fiberglass-reinforced synthetic material. The scales and the spine area near the ends of the handle are textured to improve grip. The knife also possesses a rope-cutting function with the blade closed. A forwards-facing groove in the handle is capable of cutting Paracord, fishing line, etc.

The bayonet-style clip developed by SOG (patent pending) is very cleverly designed and can be adjusted with little effort. Together with the dual thumb studs, the spring-assist and the Arc-Lock, the Trident is a capable knife suitable for use by right- and left-handers.

Thanks to the groove in the handle, when closed the blade can be used as a rope cutter.

SPECIFICATION

Manufacturer:	SOG Specialty Knives & Tools, LCC.
Model:	Trident Tiger Stripe transferred-3
Blade Length:	95.0 mm
Blade Thickness:	3.2 mm
Blade Steel:	AUS-8
Overall Length:	216.0 mm
Locking:	Arc-Lock
Weight:	102 g
Handle:	Zytel
Manufactured in:	USA
Supply Number:	none (privately procured))

FIXED KNIVES

During debriefing, Sergeant Martinez shows Staff Sergeant Perryman where his patrol had enemy contact using his Blackwater Combat Knife. Both men are members of A Company, 1st Battalion, 38th Infantry Regiment "Rock of the Marne". At the beginning of 2008 the battalion was deployed in Baqubah. Almost daily, Sunni and Shiite militias battled each other or the Iraqi security forces and the American military.

Iraq, December 2007. The American forces issue ballistic sunglasses to their soldiers to protect their eyesight. The glasses are capable of stopping small metal fragments and even shotgun pellets and can be fitted with clear or tinted lenses. This marine is wearing a model by Eye Safe Systems Incorporated (ESS). For a knife, he has chosen the Offsider combat knife by the Benchmade Knife Company.

BENCHMADE FECAS OFFSIDER

Shapely Outsider for Frontline Use

While the handle of the Offsider knife in the photo on the left is black, the fine desert dust has simply stuck everywhere. It acts like fine sandpaper, constantly abrading exterior surfaces and exposing equipment, whether metal, composite materials or fiber, to extreme wear. The Benchmade Offsider was designed by knife maker Steve Fecas, who comes from the southern USA and has been making knives since the 1970s. The Offsider is a huge combat knife, the blade alone measuring just over 18 centimeters long. Total length is an impressive 33 centimeters..

The dagger-like, modified spear point blade is up to 5.6 millimeters thick and is made of 440C rustproof steel hardened to 58 to 60 Rockwells. A feature of knives designed by Steve Fecas is their extraordinary shape, and the Offsider is no exception: the sloped butt, simulated blood channel and prominent false edge speak a clear design language. The tang, which is same thickness as the blade, extends the full length of the composite thermoplastic handle, imparting enormous strength.

The shape and surface texture of the handle ensure a secure grip, even in wet conditions. A lanyard of orange Paracord has been skillfully knotted to the handle hole of the knife in our illustration. This item is not delivered with the knife, however.

The sheath has a close-fitting insert of composite material which also secures the knife to the belt. On the outside there is a roomy pocket with Velcro flap which can accommodate a small tool, sharpening stone or other small item of equipment. Two Velcro patches on the back and an eye at the tip of the sheath enable numerous attachment options. Because of the knife's length, however, the thigh is the most logical carriage location.

The Offsider, part of Benchmade's Red Class and manufactured in the Far East, was not produced for very long. Perhaps this large and rather heavy knife (more than 350 grams) failed to touch the nerve of the military consumer – despite, or perhaps because of, its elegant appearance. A shame actually, for the Offsider impresses with numerous small details. Anyone who was able to get one already has a youthful classic.

Because of the knife's impressive length, the scabbard projects above the vest.

The production Cordura sheath can accommodate a sharpening stone.

SPECIFICATION

Manufacturer:	Benchmade Knife Company, Inc.
Model:	Fecas Offsider
Bladev Length:	181.0 mm
Blade Thickness:	5.6 mm
Blade Steel:	440C
Overall Length:	331.0 mm
Weight (with/without sheath):	–/350 g
Handle:	Thermoplastic Composite Material
Sheath:	Cordura
Manufactured in:	Taiwan
Supply Number:	none (privately procured)

1st Lieutenant Hoyle is the leader of an infantry platoon of the 1st Battalion, 38th Infantry Regiment. At the beginning of 2008 his brigade was operating in the Baqubah area of Iraq. On his ballistic combat vest, 1st Lieutenant Hoyle carries a Benchmade Nimravus.

BENCHMADE NIMRAVUS

The Small Black Knife for Tactical Use

The Nimravus model originated from the Benchmade Black Class for professional users. In the photo on the left an example of the knife can be seen in its MOLLE-compatible sheath on the lieutenant's combat vest. The Nimravus was developed for Benchmade by the well-known knife maker Allen Elishewitz and was the first fixed knife in the Benchmade line. The initial version had a G-10 handle and a blade of M2 tool steel.

With a blade length of 11.4 centimeters, the Nimravus is definitely not a full-size combat knife, and its maximum blade thickness of 2.9 millimeters is decidedly modest compared to other operational knives. Nevertheless, when it comes to wieldability and good cutting qualities, the Nimravus is hard to beat. There are two blade versions: the Model 140 BK comes with a modified Tanto blade and false edge on the spine, while the Model 141 BK has a rugged Tanto blade.

The blades of both knives are tapered, consequently the maximum blade thickness of 2.9 millimeters is retained over about a third of the blade. This reduces the already rather low starting strength. The straight edge is good for pressing and pulling cutting movements. The blades – 154-CM and D2 tool steel are used – have a BT Teflon coating by Benchmark, which reduces wear and eliminates reflection. The blades are available with partial serrations or a smooth edge.

The handle, which features vertical riffles, is now made of 6061-T6 aluminum. The tang retains its full thickness the length of the handle. The projecting tip forms a lanyard hole and can also be used as a striking aid. The nylon sheath can be attached to tactical MOLLE equipment or to a belt. A Kydex insert protects the wearer from the sharp blade. The sheath is available in a variety of colors (including ACU). A Kydex sheath and G-10 handle are also available for this handy service knife. An interesting variation is the slightly-modified Nimravus made for the company Blackwater Gear. The most important differences from the Benchmade model are the handle with Blackwater logo and the additional serrations on the blade spine.

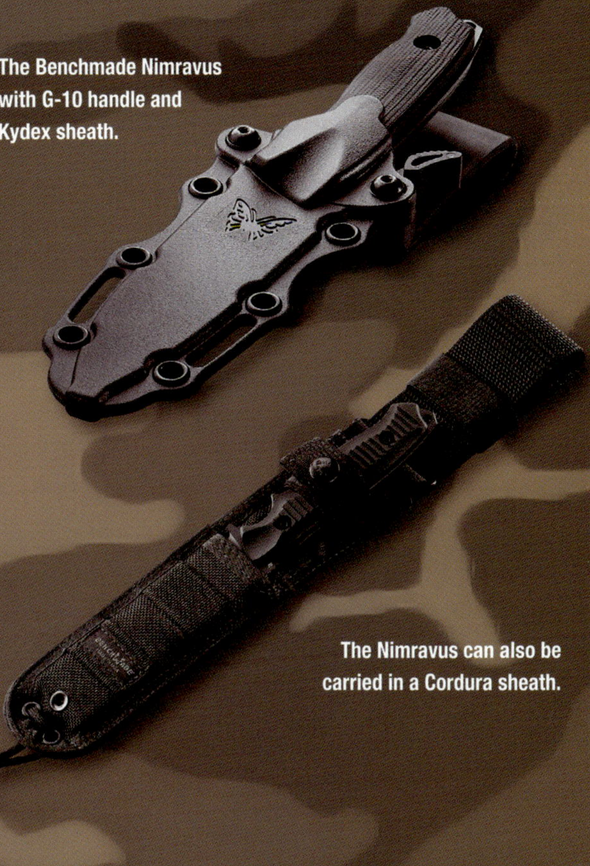

The Benchmade Nimravus with G-10 handle and Kydex sheath.

The Nimravus can also be carried in a Cordura sheath.

SPECIFICATION

Manufacturer:	Benchmade Knife Company, Inc.
Model:	Nimravus
Blade Length:	114.0 mm
Blade Thickness:	2.9 mm
Blade Steel:	154-CM or D2
Overall Length:	240.0 mm
Weight (with/without sheath):	—/—
Handle:	6061-T6 aluminum
Sheath:	Nylon or Kydex
Options:	Modified Tanto, Tanto blade, Blackwater Gear
Manufactured in:	USA
Supply Number:	NSN 1095-01-466-8569

The equipment carried by this paratrooper of the 82nd Airborne Division "All American" includes a Presidio combat knife by Benchmade. The photo was taken in March 2007 in Kadamiyah, a quarter of Baghdad. The 82nd Airborne is one of the most mobile units of the U.S. Army.

BENCHMADE FIXED PRESIDIO

Modern Operational Knife for a Multitude of Uses.

The design of the Fixed Presidio is a product of American knife-making legend Warren Osborne, a permanent member of Benchmade's design team. This heavy field and combat knife is aimed primarily at the military buyer. The modified clip point blade is made of rustproof 154-CM steel produced by the American maker Crucible Steel. The blade is hardened to 58 to 60 Rockwells.

In designing his knives, Warren Osborne places particular emphasis on the grip, and this also applies to the Fixed Presidio. The handle, made of CNC machined 6061-T6 aluminum, has systematically-machined scales, ensuring a firm grip. The volume of the handle increases towards the center, filling the hand. There is a hole at the butt end for attaching a lanyard. The handle components are screwed together and can be disassembled as required. The knife has a prominent quillion and an aggressively riffled thumb rest to prevent the hand from sliding onto the blade.

The blade, which is almost 16 centimeters long, has a thickness of 4.6 centimeters, which makes it extremely durable. The matt-black finish protects against corrosion and eliminates reflections. To increase its operational value, the Fixed Presidio has two serrated edges. One is directly in front of the handle, while the second is on the spine just behind the tip. This is an unusual position, but one seen on other knives such as the Buckmaster by Buck.

The knife comes with a multifunction soft ADC sheath produced by Eagle Industries. Inside the Cordura sheath there is a tight-fitting liner of synthetic material. A flap with button snap holds the handle to the sheath. The sheath can be attached to MOLLE vests or carried in the classic position on the upper thigh. Benchmade also equips the ADC sheath with Malice clips by Tactical Tailor and a Cordura extension to allow carriage lower than usual on the thigh, thus avoiding problems with heavy items of equipment. These are minor details, but ones which reveal the well thought-out overall concept of the knife.

The Fixed Presidio's sheath is produced by Eagle Industries, a company with a great deal of experience in this field.

The Fixed Presidio's handle is CNC-machined 6061-T6 aluminum.

SPECIFICATION

Manufacturer:	Benchmade Knife Company, Inc.
Model:	Fixed Presidio
Blade Length:	157.0 mm
Blade Thickness:	4.6 mm
Blade Steel:	154-CM
Overall Length:	285.0 mm
Weight (with/without sheath):	— /281 g
Handle:	6061-T6 aluminum
Sheath:	Cordura
Manufactured in:	USA
Supply Number:	none (privately procured)

This U.S. Marine has attached an SRK by Cold Steel to the shoulder section of his tactical vest. The man is part of a patrol from the 3rd Battalion, 23rd Marine Regiment and is armed with an M249 SAW light machine-gun. At the beginning of 2008 the battalion was responsible for security in the Iraqi city of Haditha.

COLD STEEL SURVIVAL RESCUE KNIFE

Powerful Operational Knife for Many Requirements

The textured handle offers an extremely secure grip.

For many, the SRK (Survival Rescue Knife) by the American company Cold Steel is the fixed knife of choice on account of its high quality. Its users include the U.S. Navy Seals. Until a few years ago the SRK was made in the USA, with a blade of Carbon V steel, which is fine-grained but not rustproof. Manufacture of the full-tang knife was subsequently moved to Japan, where much of the Cold Steel product line is made. The blade and tang are now made of Japanese AUS-8 steel.

When the knife was designed, care was taken to ensure that it would be rugged enough to withstand hard use. The SRK's Bowie blade is just under five millimeters thick. The grind begins about one centimeter below the blade spine, consequently the maximum thickness is retained over most of the blade. Even the tip, which is often quite thin on Bowie blades, is very sturdy. The blade has a black Teflon coating. Even under the sturdy Kraton handle the metal is coated with Teflon.

The rubber-like material of the handle provides a secure grip, even in wet conditions. A guard prevents the hand from sliding onto the blade. At the butt of the handle there is a brass-lined lanyard hole. This is useful if, for example, one wants to prevent the knife from slipping from the hand while chopping. The soldier on the left has secured his Survival Rescue Knife in several ways. He has attached it to his MOLLE loops with Paracord run through the holes in the edges of the sheath. The broad belt loop is fitted with a Velcro fastener and button snap – this double safety for the belt attachment is the result of a request by the U.S. Navy Seals. The end of the molded sheath snaps behind the hand guard, holding the knife in place. There is also a handle retention strap with snap.

Weighing 220 grams, the SRK is a rugged, versatile knife which has advantages, not just in war zones, but also in civilian use by outdoors enthusiasts.

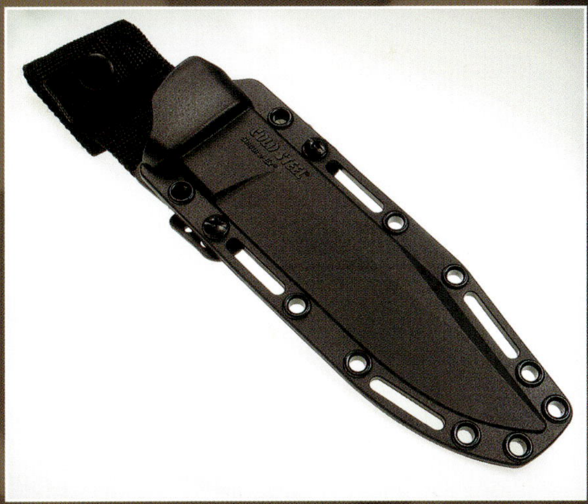

The Cold Steel SRK comes with a Kydex sheath as standard equipment.

SPECIFICATION

Manufacturer:	Cold Steel, Inc.
Model:	SRK
Blade Length:	152.0 mm
Blade Thickness:	4.8 mm
Blade Steel:	AUS-8
Overall Length:	273.0 mm
Weight (with/without sheath):	300/220 g
Handle:	Kraton
Sheath:	Secure-Ex
Manufactured in:	Japan
Supply Number:	none (privately procured)

This member of the Spanish Legion is armed with a CETME LC 5.56 x 45 mm assault rifle. In 1999 many nations, including Spain, sent elite units to Kosovo as part of KFOR, which was supposed to restore security and order there. The Spanish KFOR contingent was attached to Multinational Brigade West, led by Italy, and operated west of Pec.

COLD STEEL TERMINATOR

Compact Weapon for Combat Use

The knife the Spanish Foreign Legionnaire has attached to his ballistic vest is a push dagger made by Cold Steel. The American company Cold Steel is best known for its Tanto knives and during its existence has produced a number of now-legendary knives not limited to Japanese blade shapes: the Tai-Pan, SRK, Laredo Bowie and others. The push dagger in the photo, the Terminator, has been out of production for some time, having been replaced by the Safekeeper and Safemaker models. Most Cold Steel knives are made in Japan, but for some years the company's less expensive knives have been made in Taiwan.

The generously laid-out handle of soft Kraton rubber allows the hand to literally meld with the knife, which is part of the reason this type of knife is so dangerous. The textured surface of the handle further enhances the grip, even with wet hands. The 90-degree angle of the handle severely restricts the knife's versatility, however. Because of its design it is only suitable for stabbing, making it strictly a weapon.

According to the manufacturer, Japanese AUS-6 steel is used for the blade, with a carbon and chromium contents of approximately 0.6 percent and 14 percent respectively. Its edge-retention ability is rated as moderate (similar to 440A), but it is easy to sharpen, not particularly breakable and very rustproof. The dagger-shaped blade with narrow groove in the center is hollow-ground and comes with a finely brushed surface finish.

The knife's compact size makes it ideal for concealed carriage. The leather sheath seen in the photo is in fact not intended for carriage with the handle facing down, as the original model lacked a retention strap. The use of adhesive tape solved this problem, however. Improvisation is simply part of a soldier's life. The leather sheath was later replaced by a Kydex version.

Incidentally, in Germany push daggers such as the Terminator have been banned for some years.

Cold Steel Terminator – a pure weapon.

The first generation Terminators came with this leather sheath.

SPECIFICATION

Manufacturer:	Cold Steel, Inc.
Model:	Terminator
Blade Length:	85.0 mm
Blade Thickness:	5.0 mm
Blade Steel:	AUS-6
Overall Length:	145.0 mm
Weight (with/without sheath):	120/180 g
Handle:	Kraton
Sheath:	Leather
Manufactured in:	Japan
Supply Number:	none (privately procured)

The U.S. armed forces regularly introduced new weapons and equipment during the war in Iraq. Beginning in 2007, the 4th Battalion "Manchus", 9th Infantry Regiment tested the new infantry equipment developed by the Land Warrior Program under combat conditions. This included video reconnaissance equipment and infrared devices. This Manchu is also carrying an A.B.C. Operator's Model combat knife by CRKT.

CRKT A.B.C.'S OPERATOR'S MODEL

Suitable for Use on Land and Under Water

The A.B.C. Series by CRKT includes two knives: the Operator's Model seen in the photo with Tanto blade, and the Aqua with unsharpened blunt-tipped blade. Both knives were developed by knife maker Jim Hammond, who had been working with CRKT since 1994.

The first of the two versions was the A.B.C. Aqua. A.B.C. stands for "All Bases Covered", a reference to the knife's ability to be used on land and under water. The 9.5-centimeter blade of both versions is made of Japanese AUS-8 steel. The A.B.C. Aqua's dual-grind blade is hollow ground and has a long serrated section on the spine. With a maximum blade thickness of 4 millimeters, the A.B.C. is extremely sturdy. The blade is chisel-shaped and unsharpened to avoid damaging inflatable boats and other diving equipment and to allow the blade to be used for prying.

The Operator's Model has a dual-grind pointed Tanto blade and may also be used as a weapon in combat. It is also hollow ground, though the serrated edge on the spine is shorter than that of the Aqua Model. From the serrated edge to the tip, the spine has a false edge. The maximum thickness of the blade and tang is four millimeters.

Both knives have a black titanium nitride coating. Interestingly, the blade is polished to a high gloss before it is coated in order to make the steel more corrosion proof. The contoured handle scales screwed to the skeletonized tang and are interchangeable (black or signal yellow). Three finger choils ensure a firm grip. On the butt there is a lanyard hole. Handle length is 12 centimeters.

Thanks to a sturdy pocket clip and numerous eyeholes, the light Zytel sheath (less than 100 grams) can be attached in a variety of ways to clothing, equipment, rucksacks or diving equipment. Chicago screws allow adjustment of the knife's compression fit into the sheath to meet the user's needs.

The A.B.C. can be had with two types of blade.Klingenformen erhältlich.

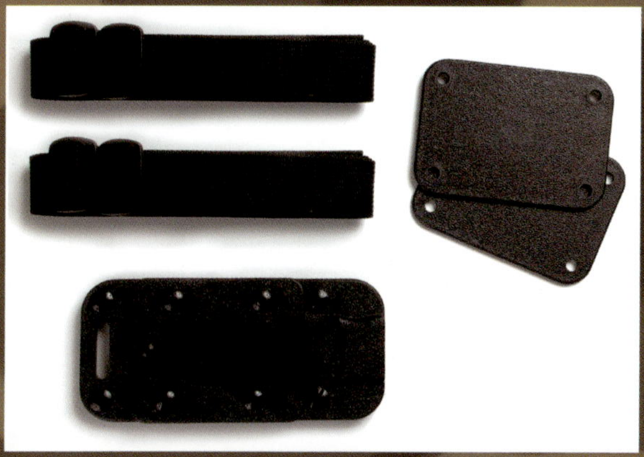

The A.B.C. sheath offers many possible attachment methods.a

SPECIFICATION

Manufacturer:	Columbia River Knife & Tool
Model:	A.B.C. Operator's Model
Blade Length:	95.0 mm
Blade Thickness:	4.0 mm
Blade Steel:	AUS-8
Overall Length:	216.0 mm
Weight (with/without sheath):	220/120 g
Handle:	Zytel
Sheath:	Zytel
Alternativen:	A.B.C. Aqua
Manufactured in:	Taiwan
Supply Number:	none (privately procured)

The first parliamentary elections in Iraq were held on 15 December 2005. In Al Karmah, north of Fallujah, Marines of G Company "Raiders", 2nd Battalion, 2nd Marine Regiment ensured quiet and order during the election. This marine is carrying a Corkum First Strike combat knife.

CRKT CORKUM FIRST STRIKE

Exotic, Sturdy Combat Knife with Cord Wrap Handle

The American sword and knife maker Steve Corkum subscribes totally to the Japanese swordsmithing tradition. He prefers making Japanese swords, Tanto, Naginata and Yari, whose blades he fabricates from non-rustproof carbon steel and selectively hardens.

Columbia River Knife & Tool has turned one of his Tanto models into a production knife. The First Strike was produced in three sizes by CRKT, with 7.9 and 11.1 centimeter blades and one with a 14.2-centimeter blade. The latter model is shown being worn on the marine's left hip. While the blade of the original version was made of selectively hardened 1050 carbon steel, the production version used Japanese AUS-6 steel hardened to about 56 Rockwells (this knife is no longer made).

The 14.2-centimeter blade has a satin finish and a maximum thickness of five millimeters (six millimeters on the Custom versions), which results in an extremely sturdy blade. The blade has a black titanium-nitride coating. The grind is convex zero edge, resulting in a very strong edge. The Tanto shape is modified rather than traditional Japanese: the tip is dropped, the edge rises at an angle.

The handle, which is more than 14 centimeters long, is made of real ray skin underlay with olive green cord wrap – perhaps not the best choice for dusty desert regions, but stylish. To save weight and maintain balance, the tang tapers toward the butt end from five to 1.5 millimeters. The end of the handle can be used as a striking point.

The injection-molded Zytel sheath can be worn on the left or right. On the back it has a removable pocket clip, which can also be attached to the sides or tip. Together with the numerous eyeholes, it offers a multitude of attachment possibilities: tip-up, tip-down, on the belt, on the equipment vest. Despite its exotic appearance, the First Strike is a thoroughly rugged and practical knife for use in the field. The only reservations concern the grip, as stylish as it is. It is simply too narrow and angular to be worked with for long periods.

The sheath (here in the standard First Strike satin finish) is made of robust Zytel synthetic material.

SPECIFICATION

Manufacturer:	Columbia River Knife & Tool
Model:	First Strike
Blade Length:	142.0 mm
Blade Thickness:	5.0 mm
Blade Steel:	AUS-6
Overall Length:	289.0 mm
Weight (with/without sheath):	430/309 g
Handle:	Ray Skin/Nylon
Sheath:	Zytel
Options:	satin-finished blade and black handle cord wrap
Manufactured in:	Taiwan
Supply Number:	none (privately procured)

Unfortunately there are no details available concerning this photo, which was provided by knife maker Allen Elishewitz. The man in the photo, however, is most likely a "private contractor", often encountered in Afghanistan and Iraq. He is armed with a Colt M4 5.56 mm x 45 assault rifle and a Beretta 9 mm x 19 pistol. The F.T.W.S. in his hand is either a handmade version by Elishewitz or an early CRKT prototype with matte blade.
Photo: Allen Elishewitz

CRKT ELISHEWITZ F.T.W.S.

Designed for Use by Those Who Serve.

The F.T.W.S. (For Those Who Serve) by the American knife maker CRKT is an outstanding knife. There are absolutely no non-essential design elements and the F.T.W.S. is uncompromisingly designed for function. The design is based on an idea by Allen Elishewitz, whose own military experience told him exactly what was required in a reliable knife.

The full tang knife is made of SK5 carbon steel, which holds an edge well and is also strong and break-resistant. As well, the steel, which is hardened to 57 to 58 Rockwells, is easily sharpened. Because of the absence of a chromium component, the steel is not rustproof. For this reason, the production version of the spear point blade is black powder coated. The knife in the photo must be an early prototype of rustproof steel, sandblasted to a matte finish.

With a maximum blade thickness of 5.8 millimeters, the F.T.W.S. is capable of enduring heavy use. Except for a few centimeters in front of the tip, the blade spine has a flat grind with a chisel ground edge. This allows the knife to be used for hacking and chopping tasks, for example. This spares the finely ground main edge. The tang extends through the entire handle, although its thickness lessens somewhat toward the butt (tapered tang), which is supposed to improve balance and grip.

The handle is made of Zytel with injection molded scales. A semi-hilt and four finger choils provide the best grip under all conditions. This is important, for with a weight of just over 300 grams the F.T.W.S. is a handful. A lanyard hole is provided at the butt end.

The sheath of the F.T.W.S. is also deserving of special mention: it is 36.5 centimeters long and 7.7 centimeters wide. Cordura and Zytel are the materials used. Dual belt loops are Velcro adjustable. There are also numerous grommets, which together with the Cordura line supplied with the knife enable it to be attached in a variety of ways. The sheath can also be attached to MOLLE loops. Two security handle straps secure the knife firmly. A utility pocket is capable of accommodating a knife, tools or sharpening accessories. As a reliable and uniquely versatile knife, the FTWS is extremely convincing.

CRKT designer and knife maker Allen Elishewitz.Allen Elishewitz.

Elishewitz incorporated an extremely strong tip in the design of the F.T.W.S.

SPECIFICATION

Manufacturer:	Columbia River Knife & Tool
Model:	F.T.W.S.
Blade Length:	160.0 mm
Blade Thickness:	5.8 mm
Blade Steel:	SK-5
Overall Length:	295.0 mm
Weight (with/without sheath):	519/306 g
Handle:	Zytel
Sheath:	Cordura/Zytel
Manufactured in:	Taiwan
Supply Number:	none (privately procured)

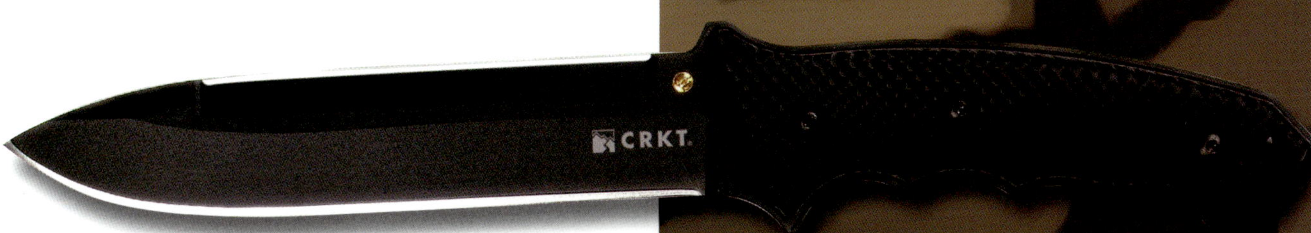

This American soldier is wearing the Advanced Combat Uniform with digital camouflage pattern. Over it is his protective gear, consisting of an Interceptor Body Armor vest with Improved Outer Tactical Vest and front and side plates plus groin protector, ballistic protective glasses, knee and elbow guards and gloves of heavy fireproof material. Combat knives, such as the M60 by CRKT seen here, are often carried on the front of the tactical vest.

CRKT M60-14 NK SOTFB

Heavy Equipment for Extreme Operations.

The SOTFB (Special Operations Tactical Fixed Blade) was developed by CKRT in cooperation with military advisors to meet the most demanding requirements. The design incorporates all of the company's practical experience. Developing the supreme cutting tool was not the primary objective, although the knife, with its several cutting edges, is capable of making extremely clean cuts. The objective was an unconditionally dependable knife for combat and field use. Full-tang construction was therefore chosen for the 269-gram knife, with the tang retaining its full thickness to the handle butt.

Maximum blade thickness is four millimeters. The flat edge is just over two centimeters long and gives way to a recurved dual-grind edge. Near the handle there is a serrated edge several centimeters long, which is useful for cutting heavy cord and nylon rope. The tip is rather more bluntly ground and is extremely robust.

The 11.1-cm Tanto blade has a titanium nitride coating to further enhance the anti-corrosion properties of the rustproof steel. The knife's center of gravity is just behind the hilt, roughly where the index finger rests. Handle length is 12.7 centimeters. As the integral hilt with thumb friction grooves and pronounced butt pommel cost some length, however, the effective grip area is only about ten centimeters – which is a little short for someone wearing heavy combat gloves. Because of the handle's strength, however, the SOTFB fits comfortably in the hand. The scales incorporated into the molded Zytel handle further contribute to a firm grip.

As the photo on the left illustrates, the sheath can be attached to MOLLE loops simply by using Paracord and the numerous grommets provided. A Zytel insert within the Cordura sheath holds the knife securely and prevents the tip from piercing the sheath. Two handle retaining straps with snaps hold the SOTFB securely in the sheath. A pocket on the front of the sheath, which can also be worn on the belt, can accommodate a folding knife, tool or sharpening equipment.

The riffled thumb rest allows controlled cutting and also prevents slipping.

The M60 SOTFB's sheath has a nylon insert to prevent piercing.

SPECIFICATION

Manufacturer:	Columbia River Knife & Tool
Model:	M60-14 NK SOTFB
Blade Length:	111.0 mm
Blade Thickness:	4.8 mm
Blade Steel:	AUS-8
Overall Length:	238.0 mm
Weight (with/without sheath):	400/269 g
Handle:	Zytel
Sheath:	Cordura/Zytel
Options:	Drop-point blade, sand color
Manufactured in:	Taiwan
Supply Number:	none (privately procured)

Coalition troops deployed in Afghanistan and Iraq must deal with extreme climatic conditions. Daytime temperatures in Iraq, for example, can exceed 50º C in summer. In the desert in winter, temperatures drop to below zero at night. Sand and dust are constant companions. This U.S. Marine has just returned from a patrol in the desert and is covered with a thick layer of dust. Note the Sting dagger by CRKT.

CRKT RUSSELL STING

A Classic with a New Look

The CRKT Sting in the photo is a little dusty – but that has nothing to do with the fact that this model first appeared on the market in the 1970s. The Sting, now again in production by CRKT, is based on a design by the legendary A.G. Russell. He is not only an innovative knife maker and founding member of the American Knifemaker's Guild, but was also the first to operate a mail order knife business, beginning in 1964. He also contributed to the establishment of the American knife magazines "Knife World" and "The American Blade" (now "Blade Magazine").

Russell developed the Sting model, a handy integral dagger which could be carried in a leather boot sheath, in the early 1970s as a hunting and self-defense knife. But no matter which manufacturer he approached, his design was rejected, all claiming that it was too expensive to make. That is until he approached Othello (Anton Wingen) in Solingen. They hesitated briefly then agreed. The first Sting daggers were sold in the USA in 1975 for 35 dollars. Today, well-preserved examples with deer horn, Micarta or wooden scales fetch prices of up to 300 dollars.

CRKT produces the model according to the old standards: 1050 carbon steel is first forged into the rough shape, then it is ground to shape and heat treated to exploit the full potential of the non-rustproof steel. From the blade spine, the blade is hollow-ground top and bottom. Maximum blade thickness is 3.5 millimeters and the dagger weighs 110 grams. A powder coat finish protects the knife against rust. The Sting's steel handle provides moderate but useful hand protection and is contoured to give a firm grip to the bare or gloved hand. At the butt end of the handle, which is 9.3 centimeters long, there is a very generous lanyard hole.

The sheath has an insert of Zytel, a tough synthetic material, which in this case serves to prevent the blade from piercing the sheath and holds the knife securely in place (it snaps into the first finger choil). The rest is made of durable Cordura material. The sheath can be attached in a variety of ways. CRKT describes the knife as one of the strongest and toughest knives the company has ever made – and the materials used and the way it is made back up this claim.

CRKT designer and knife making legend A.G. Russel.

The Sting sheath has a piercing guard of synthetic material and straps for underarm carriage.

SPECIFICATION

Manufacturer:	Columbia River Knife & Tool
Model:	Russell Sting
Blade Length:	81.0 mm
Blade Thickness:	3.5 mm
Blade Steel:	1050
Overall Length:	174.0 mm
Weight (with/without sheath):	200/110 g
Handle:	1050
Sheath:	Cordura/Zytel
Options:	Original Othello version from the 1970s
Manufactured in:	Taiwan
Supply Number:	none (privately procured)

Modern forces use helicopters for close support, to transport men and equipment, and to evacuate the wounded. In December 2004 this pilot of the 10th Mountain Division was part of the crew of a UH-60L Blackhawk transport helicopter based in Iraq. His equipment includes an ASEK I Tanto Survival Knife by the Eickhorn Company of Solingen.

EICKHORN ASEK-I-TANTO

Specialized Military Knife with Modular Construction

The ASEK-I-Tanto was designed especially for the crews of military aircraft. ASEK stands for Aircrew Survival Egress Knife.

The blade of the ASEK-I-Tanto knife is of modular construction. The actual blade is just under three millimeters thick and is strengthened by a second blade attached to the reverse side. This second blade, which is screwed to the main blade, forms the horizontal edge of the integral rope cutter. It also completes the row of teeth of the spine saw. With the saw one can cut in either direction safety glass, Plexiglas, aluminum and wood. The blade is made of 440A corrosion-resistant steel, which is also coated with black Kalgard paint. The blade's Tanto shape makes the point very sturdy. The slot for the rope and belt cutter is located several centimeters before the edge rises sharply to the tip. The other half of the blade to the cross guard is serrated.

The glass-fiber reinforced polyamide insulated handle is designed to withstand an electrical surge. Its size makes for a sure, safe grip, even when wearing gloves. At the butt end there is a lanyard hole for attaching Paracord. A steel insert reinforces the large cross guard and protects the knife hand during heavy work. Two holes make it possible to attach the knife to a shaft, forming a spear. In the butt of the handle there is a conical metal glass breaker.

The Cordura sheath has a liner of synthetic material and a MOLLE carrying system. At the bottom end of the sheath there is a water drainage hole. The knife also comes with a small push-dagger, a diamond sharpening plate, which is carried in the reverse side of the sheath, an awl and a leg belt for attaching the sheath to the calf. The ASEK-I-Tanto is undoubtedly a good choice for its actual purpose – enabling pilots to quickly escape from a crashed aircraft or rescue the injured.

The glass breaker is screwed to the butt of the handle.

ASEK push-dagger with additional screwdriver function.

SPECIFICATION

Manufacturer:	The Original Eickhorn Solingen Co. Ltd. (formerly A. Eickhorn GmbH & Co. für Schneidwaren und Waffen KG)
Model:	A.S.E.K. I Tanto
Blade Length:	135.0 mm
Blade Thickness:	4.8 mm
Blade Steel:	440A
Overall Length:	270.0 mm
Weight (with/without sheath):	500/250 g
Handle:	synthetic material
Sheath:	Cordura/synthetic
Options:	improved Aviator series with N695 steel, ABCK version by the company Waffentechnik Borkott & Eickhorn
Manufactured in:	Germany
Supply Number:	none (privately procured)

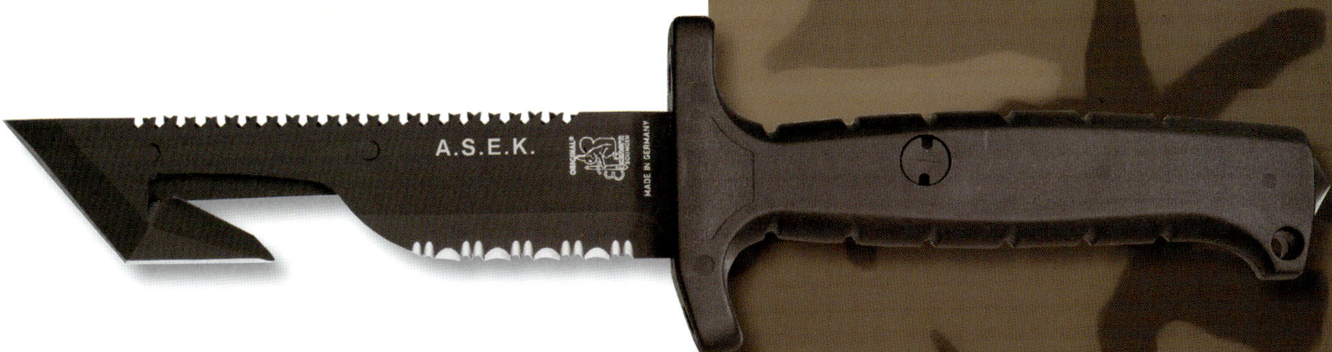

During the period from the start of "Operation Iraqi Freedom" in March 2003 to February 2009, more than 100 helicopters have been shot down by resistance fighters, destroyed in collisions with other aircraft or have crashed due to technical problems. This photo was taken during mission preparations by the crew of a UH-60L Blackhawk in December 2004. The crewman in the foreground is carrying an ASEK-II survival knife.

EICKHORN ASEK II

Escape and Survive.

The knife being worn handle-up on the MOLLE loops of the pilot's survival vest in the photo on the left is an ASEK-II which was produced by the firm A. Eickhorn GmbH & Co. for Schneidwaren und Waffen KG (now the Original Eickhorn Solingen Co. Ltd.). The ASEK series was designed specifically for the crews of NATO fixed-wing aircraft and helicopters, in order to provide them with a dependable, multifunction tool in the event of a crash in unsettled areas or the enemy rear.

The ASEK-II survival knife has a blade of 440A steel. While other steels are better at holding an edge, 440A is very break-resistant. The knife comes with a diamond sharpening plate. Also included is a small folding knife with serrated blade, belt cutter and screwdriver. A small awl also comes with the knife. The sheath has a U.S. MOLLE carrying system. Thanks to the standard leg strap, the knife can also be worn on the thigh.

The insulated handle (up to 1,000 Volts) is made of fiberglass-reinforced synthetic material, which is acid- and alkaline-resistant. At the butt end of the handle there is a removable, cone-shaped glass breaker of hardened metal. The textured handle and large cross guard contribute significantly to firmness of grip. Two spear holes in the cross guard allow the knife to be attached to a branch, for use in defense or for hunting. There is a lanyard hole in the handle end.

The 3.8-cm-thick blade retains its full thickness over much of the length and width of the blade. A black Kalgard finish eliminates reflections and protects the blade against rust. From the handle to about halfway down the blade there is an aggressive serrated edge, which cuts thick rope and synthetic materials more easily than a straight edge and remains usable for a long time. Almost the entire length of the spine is finished as a saw blade, which can be used to cut wood as well as glazed panels in aircraft.

The ASEK-II's belt cutter with integral screwdriver.

This awl is included with all ASEK models.

SPECIFICATION

Manufacturer:	The Original Eickhorn Solingen Co. Ltd. (formerly A. Eickhorn GmbH & Co. für Schneidwaren und Waffen KG)
Model:	A.S.E.K. II
Blade Length:	135.0 mm
Blade Thickness:	3.8 mm
Blade Steel:	440A
Overall Length:	280.0 mm
Weight (with/without sheath):	520/225 g
Handle:	Synthetic Material
Sheath:	Cordura
Options:	Tanto, Bowie blade
Manufactured in:	Germany
Supply Number:	NSN 1095-25-151-1584

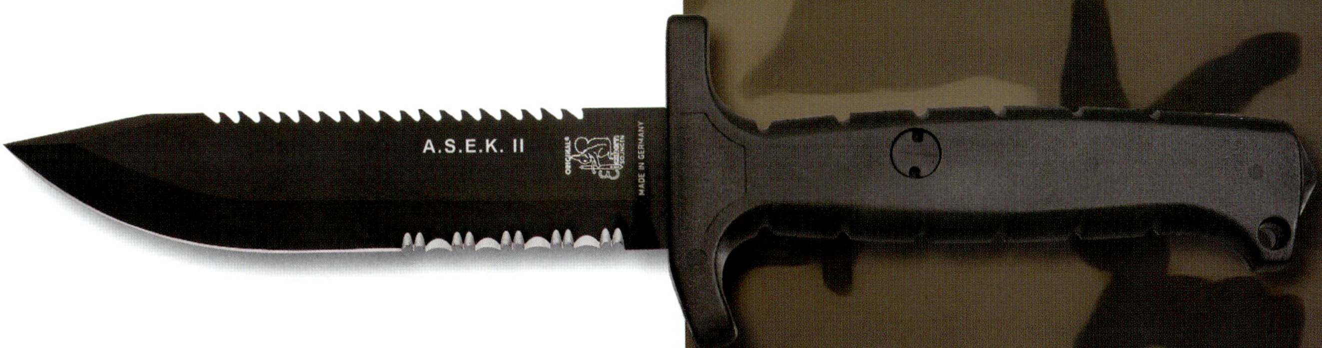

Italian special forces of the 9th Assault Regiment Col Moschin assemble at their vehicles near Bagram. They are part of the anti-terror operation Enduring Freedom (OEF) in Afghanistan. The unit was issued a combat knife by the manufacturer Extrema Ratio.

Photo: Sören Sünkler

EXTREMA RATIO COL MOSCHIN

Tactical Combat Knife for Italian Elite Units.

The Col Moschin was officially ordered by the 9th Parachute Assault Regiment "Col Moschin" in 2002 and has been in use since. The 9º Reggimento d'Assalto Paracaduisti is a special forces unit of the Italian Army. Based in Livorno, it is part of the Folgore Parachute Brigade and is also attached to the Special Forces Command (COFS) of the Italian general staff.

Each new member of this elite unit is issued a knife and trained how to use it effectively. The Col Moschin is a dagger-like knife with a straight handle like the daggers used by the Arditi, legendary Italian assault troops of the First World War. The blade does not have a central tip, however, rather it is asymmetrically formed. The blade spine has a gentle curve, placing the tip above the midpoint of the blade.

Maximum blade thickness is 6.3 millimeters. Only a few centimeters in front of the handle, the blade spine becomes a second hollow-ground edge that extends to the tip. It is ground razor sharp. Both sides of the main edge are also hollow ground. This knife geometry is ill suited to prying work, as the knife simply isn't designed for it – it is a pure weapon.

The knife's center of gravity is far back in the handle, which is intended to improve wieldability. An integral cross guard protects the hand. The aggressive serrations on the rear third of the main edge ease the cutting of tough synthetic or fibrous material. The B690 steel, made by Böhler of Austria, is considered very fine-grained and tough and is hardened to 58 HRC. The blade has a black coating.

Extrema Ratio worked with the University of Perugia to develop the optimal grip ergonomics. As a result, the handle allows many holding possibilities and rapid grip changes. The end of the projecting tang can be used to attach a lanyard and also as a striking aid. The Col Moschin is held in the sheath of fiberglass-reinforced synthetic material in two ways: by a pressure mechanism inside the sheath and by a strap with snap. Elastic cord for attaching the knife to the thigh is included with the knife.

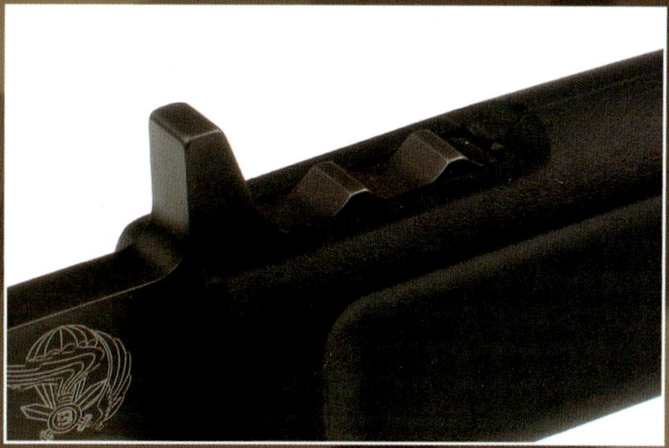

On the Col Moschin the tang also acts as a hand guard.

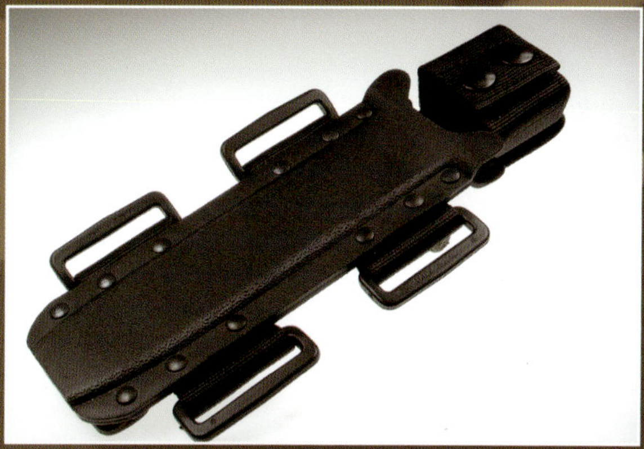

The Col Moschin sheath of Kydex is elaborately-made and offers a variety of attachment possibilities.

SPECIFICATION

Manufacturer:	Extrema Ratio Knives Design
Model:	Col Moschin
Blade Length:	156.0 mm
Blade Thickness:	6.3 mm
Blade Steel:	N690
Overall Length:	288.0 mm
Weight (with/without sheath):	420/260 g
Handle:	Forprene
Sheath:	Kydex
Alternative:	Chisel grind, civilian version
Manufactured in:	Italy
Supply Number:	none

The paratroopers of the 82nd Airborne Division "All American" stationed at Fort Bragg are members of one of the U.S. Army's elite units. In 2009 brigades from the division were stationed in Iraq and Afghanistan. At the beginning of 2009 the 3rd Brigade Combat Team was operating in the eastern part of Baghdad. This sergeant is a member of C Company, 2nd Battalion, 505th Parachute Infantry Regiment. He is carrying a Gerber Guardian.

GERBER GUARDIAN BACK UP

Gerber Classic from the Eighties.

At the beginning of the Eighties the Guardian was advertised as "A lightweight survival knife that's made to survive – the Guardian by Gerber". The original Guardian was based on a design by American knife maker Bob Loveless. The numerous model variations, including the Guardian Back Up, all bear a strong resemblance to the original model. The Guardian often turns up as a defense or combat knife, in real life and in films such as "Tomb Raider" and "Splinter Cell".

At 8.6 centimeters, the blade of the Guardian Back Up is somewhat shorter than that of the original Guardian with aluminum handle, and the sheath is made of a tough synthetic material instead of leather. The knife's sheath system is well thought-out and made. Two adjustable bolts at the mouth of the sheath allow draw resistance to be set. The Guardian Back Up can also be carried safely with the handle down. The reversible and adjustable clip can be attached to bootlegs, belts and equipment straps. A grommet with a diameter of four millimeters at the tip of the sheath enables a lanyard of 550 Paracord to be attached or a chain for use as a neck knife.

The Guardian Back Up's handle is made of a fiberglass-reinforced synthetic material with rubber-like Santoprene inserts, which significantly improve grip. This is an advantage not to be underestimated, especially with wet hands. The tang only extends about three-quarters of the way through the handle at full thickness. A pronounced pommel at the end of the handle and the shallow cross guard keep the knife from slipping from the hand.

The blade of "High Carbon Steel" (Gerber is rarely more specific in identifying steel types) has a black coating. The sharp-ground, central-tip dagger blade make this excellent knife a weapon not to be underestimated. This blade shape is far from ideal for everyday use, as the danger of self-inflicted injury is quite high. As well, the user cannot exert pressure on the blade spine. The blade is relatively thin toward the tip, making it liable to break under bending loads. Because of its light weight of less then 90 grams and its multiple attachment possibilities, however, the Back Up dagger is a good choice for an emergency knife.

The Guardian models from the 1970s were equipped with aluminum handles.

The Gerber Guardian Back Up's plastic sheath has an adjustable clip to allow the knife to be carried tip-up or tip-down.

SPECIFICATION

Manufacturer:	Gerber Legendary Blades
Model:	Guardian Back Up
Blade Length:	86.0 mm
Blade Thickness:	3.0 mm
Blade Steel:	400
Overall Length:	183,0 mm
Weight (with/without sheath):	87/50 g
Handle:	GFK/Santoprene
Sheath:	Synthetic material
Manufactured in:	USA
Supply Number:	none (privately procured)

The majority of the units of the Multinational Division South East (MND SE) in Iraq were made up of British troops. Operations in Iraq by British forces officially ended on 30 April 2009. 1st Battalion, The Royal Regiment of Fusiliers, part of the 7th Armoured Brigade stationed in Germany, carried out three six-month deployments to Iraq between 2003 and 2009. This member of the battalion was photographed in July 2008 during a patrol whose mission was to guard Basra airport against attacks. On his tactical vest is a Gerber LMF II.

GERBER LMF II

An Extremely Durable and Multifunctional Knife

The soldier at left has chosen an LMF II survival knife by Gerber. Few parallels to its predecessor, the popular LMF, are discernable. The LMF II was developed to specifications by Jeff Freemann, a longtime employee of the U.S. Army (including as a mechanic and took maker).

The LMF II is capable of withstanding the toughest demands and is extremely versatile. The 12.3 centimeter-long coated blade is made of Swedish 12C27 steel, which if tempered properly is capable of holding an edge despite its relatively high chromium content of 13.5 percent. The steel is also extremely rust-resistant. Blade thickness is a hefty 4.7 millimeters. The drop point blade is ground on both sides and has a prominent belly to the tip. Almost half of the cutting edge has aggressive serrations.

The full tang extends the length of the fiberglass-reinforced synthetic handle, which is insulated and resistant to acidic and alkaline substances. The handle, which is partly flattened with recesses, can be attached to a shaft using the three holes and some cord – allowing the LMF II to be used as a spear. A cross guard and the textured Soft Grip outer surface of the handle ensure a firm grip.

A special feature of this knife is its handle butt of hardened steel. It is not connected to the tang, in order to cushion blows. A flat hammer head makes it possible to drive nails, while the pointed tip is capable of breaking simple glass panels and even Plexiglas windows in helicopters.

The knife rests securely in the flat sheath of nylon and fiberglass-reinforced synthetic material. The sheath is equipped with a fixed sharpening attachment. Two handle straps hold the LMF II in the sheath. The LMF II can be carried on the belt or tied to the thigh using straps. Several holes allow alternate attachment possibilities, for example to equipment using zip ties. For several years, an ASEK (Aircrew Survival Egress Knife) variant of the LMF II has also been available. It comes with a belt and rope cutter which can be carried in a pouch.

The sheath also acts as a sharpener.

Holes in the cross guard make it possible to turn the knife into a spear.

SPECIFICATION

Manufacturer:	Gerber Legendary Blades
Model:	LMF II
Blade Length:	123,0 mm
Blade Thickness:	4.7 mm
Blade Steel:	12C27
Overall Length:	269.0 mm
Weight (with/without sheath):	660/325 g
Handle:	synthetic material
Sheath:	nylon/synthetic material
Options:	Desert and Foliage Green colors, ASEK
Manufactured in:	USA
Supply Number:	none (privately procured)

Afghanistan, Shahikot Valley, March 2002: during Operation Anaconda, a machine-gun team of the 1st Battalion, 187th Infantry Regiment of the 101st Airborne Division keeps a look-out for hidden Taliban fighters. The soldiers are wearing a mix of BDU desert uniforms and Woodland camouflage equipment, which was then standard. The soldier on the right is carrying a Gerber Mark II on his tactical vest. The sheath reveals that it is an early production knife.
Photo: Visier/U.S. D.O.D

GERBER MARK II

Used by American Soldiers Worldwide Since the Vietnam War

Gerber developed a desert camouflage version of the Mark II specially for use in Afghanistan and Iraq. The sheath and the handle are finished in digital camouflage pattern.

Even today, the special units of the U.S. Army appreciate fighting knives with style and history. The knife the soldier in the photo is carrying is a Mark II by Gerber. The knife was designed by Captain C.A. "Bud" Holzman. The first prototypes (eight were made in 1966) were tested by members of the U.S. Army's Fourth Infantry Division at Fort Lewis, Washington. Production of the improved model began in October 1966, and deliveries of the Model 1001 began in early 1967. The Mark II became a very popular privately-purchased knife, especially among American troops in Vietnam.

By the time production was temporarily halted in 1999, there had been numerous changes. Until 1969 the knife was sold as the Mark II Combat, then the name was changed to Mark II Survival. Non-rustproof L6 tool steel was used for the blade until 1980. In addition to 0.7 percent carbon, this steel included small quantities of chromium (up to 1.2%) and manganese (0.8%) plus molybdenum and nickel. This composition made the steel fine-grained and break-resistant.

In 1980 the L6 steel was replaced by 440C, then from 1997 to 2000 440A was used. The present-day version uses 420HC steel and is black-coated.

A feature of Mark IIs made in the first two years was the five degree angle between blade and handle, which was supposed to keep the knife closer to the body. In 1971 a relatively fine serrated edge was added. From 1977 a heavier, more aggressive pattern was used. A feature of Mark II knives until 1980 was the tailing of the blade – or recurve in today's parlance.

There have also been changes in the handle and sheath: originally the knife had an extremely rough, black-gray aluminum handle, which was later replaced by a black Armorhide handle. Now and then, however, one sees a Mark II with an orange- or yellow-painted handle: these were used as diving knives. Leather (initially brown, later black) was used for the sheath until 1986, when it was replaced by Cordura. The current version of the Mark II is still a good combat knife, though it lacks the charm of the old Mark II knife from the days of the Vietnam War.

SPECIFICATION

Manufacturer:	Gerber Legendary Blades
Model:	Mark II
Blade Length:	172.0 mm
Blade Thickness:	6.4 mm
Blade Steel:	420HC
Overall Length:	405.0 mm
Weight (with/without sheath):	–/–
Handle:	Aluminum
Sheath:	Cordura
Options:	collector's version, desert camouflage version
Manufactured in:	USA
Supply Number:	none (privately procured)

Gurkhas are masters in the use of the modern assault rifle and the Kukri knife. Gurkhas of the 2nd Battalion, The Royal Gurkha Rifles (2 RGR) carried Kukris during their deployment to Afghanistan in 2005. They were twice as long as the knives usually used by soldiers of the Gurkha regiments today. This Gurkha officer was photographed near Mazar-i-Scharif in the summer of 2005.

GURKHA CHAINPURE

Traditional Kukri from the Time of the Kings

Since the beginning of the Nineteenth Century, Gurkha soldiers have fought beside the British and Indians in almost every theater of war: during the Indian Revolt in 1857, in the trenches of Gallipoli during the First World War, in North Africa and at Monte Cassino during the Second World War, against Argentina in the Falkland Islands in 1982, as part of the IFOR peacekeeping force in the former Yugoslavia in the 1990s, in Afghanistan as part of the ISAF operation, and of course also in Iraq.

The Kukri being carried by the soldier in the photo is a so-called Chainpure. Knives with this blade length are often used in Nepal for butchering goats. The Chainpure can be identified by its very characteristic handle with a small keeper to secure the tang at the butt section, a very narrow blade and a hole (Cho) in the blade near the handle – a symbol of the god Shiva, which is supposed to give the knife holy power in battle.

As the Gurkhas have always been free to choose which Kukri they carry into action, and the Nepalis generally like everything shiny, this soldier has selected a Chainpure with a splendid decorated sheath. The Chainpure is usually carried in a traditional sheath of leather and wood. The example this soldier is wearing is an ornate sheath. It is made of wood and metal and has inlaid national insignia from the days when Nepal had a king: these include the royal crown, crossed Kukris, a fish and more.

The Kukri is part of the traditional dress of the male Nepalese, who often receive their first Kukri when just boys. They perfect the handling of the big knife and universal tool in years of daily practice. In Nepal the Kukri (also spelled Khukri or Khukuri) is used in food preparation, for opening canned food and boxes, to cut firewood and to slash paths through undergrowth.

The Chainpure is hand-forged in Nepal. The Kukri model British Service No. 1 is made of carbon steel taken from the leaf springs of recycled truck suspensions. The blade spine has an impressive thickness of ten millimeters. The spine and body of the blade are not as hardened as the edge area in order to give the blade flexibility. It is almost impossible to destroy the blade of this size of Kukri using normal means.

Typical feature of the Chainpure: the small keeper on the butt section of the handle..

Cho, the hole in the blade, symbolizes the eye of Shiva.

SPECIFICATION

Manufacturer:	Nepalese Smiths
Model:	Chainpure
Blade Length:	390.0 mm
Blade Thickness:	10.0 mm
Blade Steel:	Leaf spring steel
Overall Length:	550.0 mm
Weight (with/without sheath):	1,100/700 g
Handle:	wood
Sheath:	wood/metal
Manufactured in:	Nepal
Supply Number:	none (privately procured)

The British Army includes two Gurkha infantry battalions. The 1st Battalion, The Royal Gurkha Rifles (1 RGR) and the 2nd Battalion, The Royal Gurkha Rifles (2 RGR). Stationed in Brunei, the 2 RGR is the only unit that permanently practices jungle fighting. The soldier in the photo is a machine-gunner of the 2 RGR in northern Afghanistan in the summer of 2005.

GURKKA BRITISH SERVICE NO.1 KUKRI

Powerful Elite Fighting Knife with a Long Tradition.

After a military conflict between the British-Indian coalition and Nepal (then under the reign of the Gurkha king Prithvia Narayan) which ended in favor of the British, Great Britain was given the right to recruit mercenaries from the Nepalese people to serve in their and the Indian armies. Since 1814 Gurkha troops have served in almost every theater of war in the world.

The Kukri, which the people of Nepal have used as an all-purpose tool and weapon since at least the middle of the 17th Century, is part of the Gurkha soldier's standard equipment. The Kukri Service No. 1 of the British Army Gurkhas is the most commonly encountered version. A somewhat longer and broader blade was standard until the 1950s. With a blade about 26 centimeters long, the British Service No. 1 is relatively wieldy as Kukris go.

It is manufactured in East Nepal and is made of typical Kukri materials: non-rustproof steel from recycled leaf springs is used for the forged curved blade. Selective tempering of the blade (the edge area is more highly tempered than the body of the blade) and the considerable spine thickness of up to ten centimeters make the Kukri a serious household knife, tool and even weapon.

Typical features of the Kukri are its downward curved blade, the widening of the blade beyond this curve, and the small portion of tang which extends through the butt of the handle and is then hammered flat. The knife's center of gravity is relatively far forward. Because of this and the forward-dropped blade edge, the Kukri is particularly well-suited to chopping.

The handle of the Service No. 1 is made of water buffalo horn, which is glued to the tang. The glue used is tree resin (Laha). The sheath (Chape) is made of leather-covered wood. Also typical of Kukris are the two small knife tools accommodated in the sheath: the Chakma, a small steel for sharpening and striking sparks, and the Karda, a small knife for cutting. The British Service No. 1 is a true all-purpose tool and in practiced hands it is also a formidable offensive and defensive weapon.

There is a metal insert where the blade and handle meet, although this provides nothing in the way of hand protection.

The traditional Kukri sheath with two small knives.

SPECIFICATION

Manufacturer:	Nepalese Smiths
Model:	British Service No. 1
Blade Length:	265.0 mm
Blade Thickness:	9.0 mm
Blade Steel:	Leaf Spring Steel
Overall Length:	375.0 mm
Weight (with/without sheath):	640/440 g
Handle:	Water Buffalo Horn
Sheath:	Wood/Leather
Manufactured in:	Nepal
Supply Number:	none (privately procured)

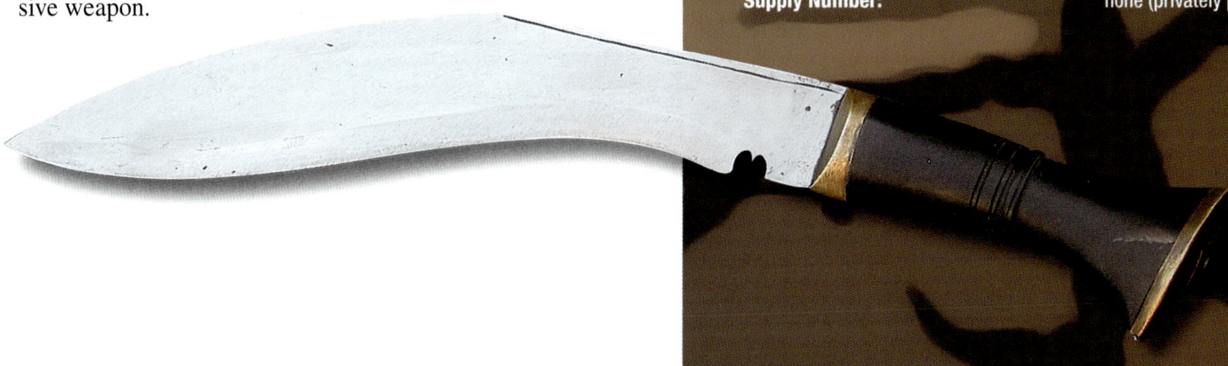

Sergeant Wagner is carrying a BK-7 combat knife by Becker Knife and Tool. He is armed with an M4A1 5.56-mm carbine fitted with a Trijicon ACOG sight. The M4A1 is part of the M16 family and measures 854 mm in length with the shoulder stock extended. In recent years it has replaced the 1,000-mm-long M16 assault rifle as the standard weapon of the U.S. Army's infantry units.M16 Sturmgewehr als Standardwaffe ersetzt.

KA-BAR BECKER KNIFE & TOOL COMBAT UTILITY KNIFE CM-BK7

Modern Combat Knife with Good All-Round Characteristics

Scarcely noticeable next to the radio: a BK-7 in the outer pocket of a tactical vest.

The BK Combat Utility Knife is a modern classic. It was one of a total of nine models developed by Ethan Becker (trained chef, outdoor professional, former president of Colorado Mountain Industries) and produced in cooperation with knife maker Camillus. Camillus went bankrupt in 2007, however Becker was able to retain control of his designs. Since 2009 his knives have been produced by Ka-Bar.

The BK-7 was developed for use by soldiers, not just as an emergency weapon, but as a knife for daily use as well. There are several parallels with the MK-2 combat knife produced by Camillus during the Second World War. The BK-7's clip point blade is almost 18 centimeters long. Just under the 4.8-millimeter-thick spine, the knife has a flat grind, which results in a blade geometry that cuts well but is also capable of withstanding heavy use, as in chopping and prying.

The low-alloyed, non-rustproof 0170-6C steel was developed by Sharon Steel, which no longer exists. It contains about one percent carbon and small percentages of chromium and vanadium. Cold Steel uses a very similar steel with the designation Carbon V. The steel has a fine texture and is considered tough and capable of holding an edge. It is a good choice for a utility blade. The blade has a black epoxy powder coating for rust protection.

The tang extends the full length of the handle. A small section projects from the butt, and this can be used as a striking aid. The handle is made of GV6H, a heavy-duty, fiberglass-reinforced synthetic material. The prominent front and rear quillions provide a firm grip.

In contrast to other BK&T models, the BK-7 does not come with a pure Kydex sheath, but instead with a nylon sheath with Kydex insert. A front pocket accommodates additional items of equipment (Multi-Tool, sharpener, etc.). The sheath can be mounted on the belt or, using the ALICE Clip, can be carried on the LBE tactical vest. The BK-7 Combat Utility is a pleasing survival and combat knife with good all-round characteristics.

The BK-7 sheath has a front pocket for tools or a sharpener.

SPECIFICATION

Manufacturer:	Ka-Bar Knives, Inc.
Model:	CM-BK7
Blade Length:	178.0 mm
Blade Thickness:	4.8 mm
Blade Steel:	010-6-C
Overall Length:	318.0 mm
Weight (with/without sheath):	500/370 g
Handle:	GV6H
Sheath:	Nylon
Manufactured in:	USA
Supply Number:	none (privately procured)

In 2009 the new Iraqi Army consisted of 260,000 men organized into 16 divisions with a total of 105 combat battalions. This soldier of the 14th Division was photographed in Basra in October 2008. His job was to guard a bridge over the Schatt-el-Arab. On his belt he wears a Ka-Bar Black Tanto.

KA-BAR BLACK TANTO

Modern Fighting Knife with a Japanese Influence

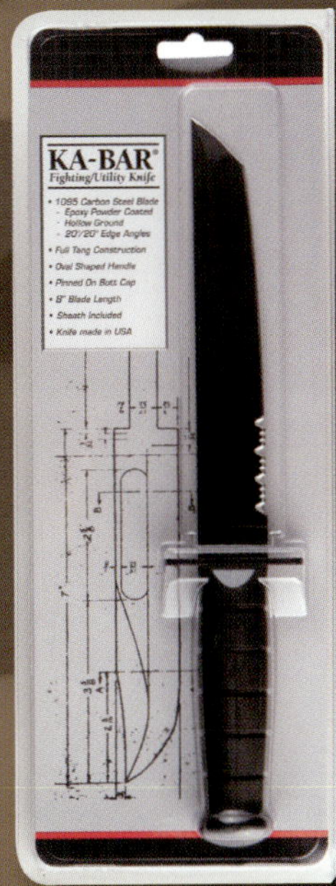

In this Spartan plastic packaging, the Black Tanto is available to U.S. military personnel in many specialized shops.

The American company Ka-Bar has been providing soldiers with field knives since 1942. The knife on the Iraqi soldier's belt displays classic design features. Experience with the 1095 carbon steel, which Ka-Bar tempers to about 57 Rockwells, has quite obviously been good, for it is used in almost all their military knives. With a content of just under one percent carbon and very small quantities of other elements (0.5% sulfur, 0.4% phosphorous, 0.8% manganese), the steel is very fine-grained and quite highly hardenable (up to 61 HRC). Ka-Bar does not fully exploit this potential, however, in favor of tensile strength and edge-holding ability.

The Tanto blade retains its maximum thickness of 4.2 millimeters over much of the blade and is thus very sturdy. Even a certain amount of misuse as a chisel or step will not bend a blade of this type. The blade is hollow-ground on both sides, with a grind angle of 20 degrees on each side. This produces a sturdy but relatively cut-friendly edge. In addition to a smooth edge there is a version with a partially serrated edge. Although the blade spine has a false edge extending back several centimeters from the tip, the tip is strong enough to easily pierce even hard materials like sheet metal or aluminum without causing damage.

Instead of the leather grip seen on many Ka-Bar knives, the Black Tanto's handle is made of Kraton thermo-elastic material. Its oval shape and five recessed rings are features seen on older Ka-Bar models, such as the USMC model. This grip design has proved very effective, as it fills the hand and is slip proof. A prominent cross guard prevents the hand from sliding forward. The butt cap is made of powdered 1095 steel and can be used for hammering.

A plastic sheath holds the knife securely and has a nylon belt loop. Numerous holes and cutouts at the edge of the sheath provide other attachment options. A handle strap and another strap over the cross guard (both with snaps) can also be used to hold the knife in the sheath.

The Kydex thermoplastic sheath holds the knife with the aid of two retaining straps.

SPECIFICATION

Manufacturer:	Ka-Bar Knives, Inc.
Model:	Black Tanto
Blade Length:	134.0 mm
Blade Thickness:	4.2 mm
Blade Steel:	1095
Overall Length:	238.0 mm
Weight (with/without sheath):	–/170 g
Handle:	Kraton
Sheath:	Kydex
Manufactured in:	USA
Supply Number:	none (privately procured)

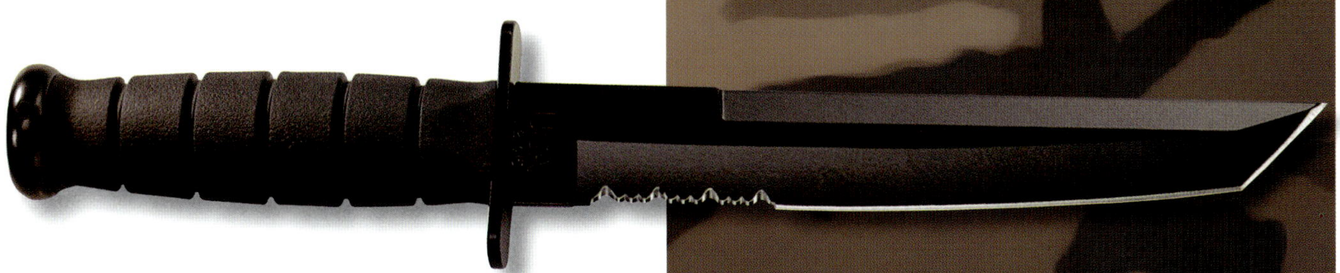

Specialist Zell has attached a Ka-Bar Large TDI to the loop system of his tactical vest. He is part of the crew of an M1126 Stryker ICV armored combat vehicle of the 1st Battalion, 38th Infantry Regiment of the 4th Stryker Brigade Combat Team of the 2nd Infantry Division. The Stryker Brigade Combat Teams are equipped with eight-wheel-drive armored combat vehicles.

KA-BAR LARGE TANTO TDI KNIFE

Backup Knife for Professionals

One size smaller: the Small TDI with coyote brown handle.

The knife attached to MOLLE loops in the photo on the left is the largest model in the TDI series, which was developed by the American Tactical Defense Institute in cooperation with John Brenner. Originally, the series was specially designed as a self-defense knife for police officers. The smaller models could be carried beneath the police belt – preferably opposite the pistol holster in the forward hip area – or on the lower leg. Typical for the TDI series is the angle between the handle and the blade – the knife is grasped and drawn like a pistol. Many policemen know this movement by heart and can execute it safely even in stressful situations.

The Large Tanto was designed primarily for soldiers, who can carry the knife openly on their equipment. A plastic sheath holds the knife securely. Cordura loops can be attached to the back side of the plastic sheath, and these allow it to be secured to the belt or MOLLE loops. The blade is 9.4 centimeters long and retains its maximum blade thickness of three millimeters over much of the length and three-quarters of the width of the blade. The grind is hollow, the edge is ground at an angle of 20 degrees (40 degrees total). The blade is made of Japanese AUS-8 steel with a Rockwell hardness of 57 to 59. The steel is considered rustproof, nevertheless it has a relatively rough non-reflective black powder coating.

The tang extends the full length of the handle. The textured Zytel handle scales are held in place by Torx screws. The grip really is exceptionally comfortable and secure. This is enhanced by the riffling at the junction between the blade spine and front of the handle and on the bottom end of the handle. Even if the knife was designed for self-defense purposes, it can be used for many other tasks. Even the angle between the blade and handle is no hindrance – for example when using the knife in the kitchen to slice onions. The knife is also quite rugged.

The bands on the back also allow the sheath to be attached to MOLLE vests.

SPECIFICATION

Manufacturer:	Ka-Bar Knives, Inc.
Model:	Large TDI Tanto
Blade Length:	94.0 mm
Blade Thickness:	3.0 mm
Blade Steel:	AUS-8
Overall Length:	192.0 mm
Weight (with/without sheath):	–/– g
Handle:	Zytel
Sheath:	Plastic
Options:	drop-point, various sizes, scales and sheath in a variety of colors
Manufactured in:	Taiwan
Supply Number:	none (privately procured)

"The Ka-Bar" has been standard equipment of the U.S. Marines since the Second World War. It still enjoys great popularity today. This gunner of the 1st Battery, 3rd Battalion, 11th Marine Regiment is carrying a Ka-Bar USMC. He is part of the crew of a 155-mm M777A2 field howitzer and was photographed carrying a high-explosive round to the gun.

KA-BAR USMC FIGHTING KNIFE

In Use Since 1942, the USMC Is One of the Most Famous of All Fighting Knives.

Soon after the United States entered the Second World War, the Union Cutlery Company (the name Ka-Bar was not officially adopted until 1952) began delivering a military knife designated the Ka-Bar. The knife had been designed in cooperation with the Marine Quartermaster Department to meet the needs of the troops, who were dissatisfied with the M1917 bayonets and Mark I daggers then in use.

The design of the "United States Marine Corps" knife goes back to a hunting knife produced since the end of the Nineteenth Century. The blade was lengthened somewhat for the Marines and was given a fuller ("blood groove"). There is a cross guard at the front of the handle, and a steel butt plate is riveted in place. The demand for the USMC knife was so great that the U.S. government issued contracts to other companies to make the model, including the Ontario Knife Company, Camillus Cutlery and Case – however the original came from Ka-Bar. That company alone produced more than a million USMC knives during the war. The legendary fighting knife is still made today in Oleans, USA, as can be seen from the blade stamping.

The design, which has remained unchanged for more than 65 years, is still justified today. Almost 18 centimeters long, the clip point blade is made of 1095 steel (Rockwell hardness 56-58). While not particularly rustproof, it is fine-grained, possesses great tensile strength and is easily sharpened. To prevent rusting, the blade, handle and pommel are given a powder-epoxy coating. The flat grind begins two-thirds of the way down the blade. With a thickness of 4.3 millimeters, the blade is very sturdy, allowing the knife to be used for light levering work and opening cans. The steel pommel can be used to drive tent stakes.

The Marine Combat Knife is a true all-round knife. The oval handle of compressed leather discs feels good and rests comfortably and securely in the hand. The USMC is a classic that continues to do its duty even today – not just with the Marines but in other services as well.

A Kydex sheath is optional.

The riveted leather sheath with USMC stamping is standard equipment.

SPECIFICATION

Manufacturer:	Ka-Bar Knives, Inc.
Model:	USMC
Blade Length:	178.0 mm
Blade Thickness:	4.3 mm
Blade Steel:	1095
Overall Length:	302.0 mm
Weight (with/without sheath):	420/300 g
Handle:	Leather
Sheath:	Leather
Options:	serrated blade, various special issues
Manufactured in:	USA
Supply Number:	NSN 1095-00-392-4102

For a long time Al-Qaida controlled the rural region southwest of Baqubah. The population fled to escape the terrorists or were forced to lend their support. The area was also used as a base of operations by the terrorists. In January 2008 the Iraqi Army launched an offensive in the region. In bitter fighting, mastery of the area was wrested from Al-Qaida town by town. Here Iraqi soldiers and American troops of the 2nd Battalion, 12th Field Artillery Regiment guard American trucks bringing supplies to the Iraqi troops. The Iraqi soldier in the foreground is carrying a Next Generation Fighting Knife by Ka-Bar on the calf of his right leg.

KA-BAR NEXT GENERATION

The New Version of the Original USMC Knife

In the mid-1990s Ka-Bar released the fully-redesigned next generation of the USMC classic. Instead of the 1095 carbon steel often used by Ka-Bar, the blade is made of powder-coated 440A steel. This steel contains about 0.3% less carbon than 1095 (or about 0.65%) but has additional elements such as 17% chromium, for example. Together with a protective coating, this ensures a high level of corrosion resistance, resulting in a blade that requires little maintenance. 440A is easy to sharpen and does not break easily, even under leverage loads.

The 17-centimeter-long Bowie blade is 4.3 millimeters thick and is flat ground on both sides. The blade's total grind angle is 40 degrees – sturdy yet excellent for cutting. There are fullers on both sides above the ground edge, making the knife somewhat lighter and more handle-heavy.

The design of the handle is also completely new: instead of leather, the handle material is Kraton, a thermoplastic elastomer. This material offers a convincingly-secure grip, even in wet conditions. Five grooves in the handle, the butt and hand guard prevent the hand from sliding. The pommel and hand guard are also made of 440A steel.

Ka-Bar offered a Kydex sheath with the Next Generation, becoming one of the first American manufacturers to do so. The sheath is made by Blade Tech. The knife rests firmly in the sheath without straps and can be drawn quickly. For rappelling or parachuting there is an additional rubber retaining band at the top of the sheath. Production of this innovative Kydex sheath has since been halted, however. Instead the Next Generation is offered with a simplified plastic sheath or alternatively with a Cordura or leather sheath. While the leather sheath is only designed for classic carriage on the belt, the Cordura version with tip guard can be attached to Load Bearing Gear or a rucksack.

For customers desiring a steel with better edge-retention characteristics, the Next Generation is also offered as the Extreme Fighting Knife with a D2 steel blade (1.5% carbon and about 12% chromium) and a pommel of rustproof 1095 steel.

The Ka-Bar Next Generation with Cordura sheath.

The D2 Extreme version of the Next Generation.

SPECIFICATION

Manufacturer:	Ka-Bar Knives, Inc.
Model:	Next Generation
Blade Length:	178.0 mm
Blade Thickness:	4.3 mm
Blade Steel:	440A
Overall Length:	302.0 mm
Weight (with/without sheath):	–/354 g
Handle:	Kraton
Sheath:	Plastic
Options:	serrated edge, with D2 blade
Manufactured in:	USA
Supply Number:	none (privately procured)

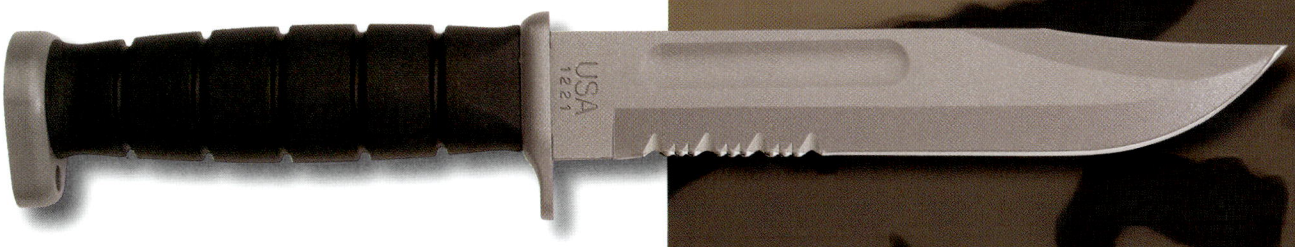

The MV-22 Osprey tilting-rotor transport aircraft made its operational debut during "Operation Iraqi Freedom". Beginning in October 2007, ten MV-22 Ospreys were stationed in western Iraq. Captain Elizabeth A Okareeh-Baah was one of the first women trained to fly the MV-22. Her equipment includes a Ka-Bar Warthog. The photo was taken at Al-Aasad airfield in January 2008.

KA-BAR WARTHOG

Combat Knife with a Massive Blade

Ka-Bar seems to have given much thought to the naming of this model (which is no longer in production). Like its namesake the Warthog, the knife is no beauty. The extremely wide and curved blade and the broad cross guard give the knife an almost bizarre appearance. The blade is too broad for fine cutting, for chopping the heavy blade could be somewhat longer, and the cross guard interferes with digging. And yet the Warthog is not all that unsuitable for use as a field knife for cutting ropes, hoses and cables, cutting wood, opening rations cans and boxes and, in an emergency, even for self-defense.

The knife is well-suited for cutting every conceivable kind of material. The relatively short but broad blade is also excellent for chopping. The non-rustproof 1095 steel (with powder-epoxy coating) is fine-grained and tough, which means that it endures mistreatment (dropping, bending loads) well. As well, the blade can be sharpened using simple means.

The broad blade gives way to a flat tang extending into the Kraton handle, to which is riveted a pommel of 1095 steel. The very wide cross guard prevents the hand from slipping onto the blade. The handle grooves and rubberlike material enhance the grip considerably and permit fatigue-free work. A good grip is essential – and the Warthog definitely provides this.

A well-fitting, penetration-proof sheath also contributes significantly to the usability of a fixed knife. The Warthog's sheath consists of fiberglass-reinforced plastic and a nylon belt attachment. Two straps, one for the handle the other for the crossguard (both with snap closure) hold the knife in place. The numerous grommets also permit other carriage configurations. The pilot in the photo, for example, has used two zip ties to attach the sheath to the horizontal loops on her vest. This carriage style with watch is rather unusual, but suggests that this pilot sees her knife essentially as a tool and not a fighting knife that must be available instantly in an emergency.

Thanks to its numerous slots and grommets, the sheath offers a multitude of carriage possibilities.

The Warthog Folder's design is based on that of the fixed-blade model.

SPECIFICATION

Manufacturer:	Ka-Bar Knives, Inc.
Model:	Warthog
Blade Length:	137.0 mm
Blade Thickness:	4.3 mm
Blade Steel:	1095
Overall Length:	257.0 mm
Weight (with/without sheath):	460/354 g
Handle:	Kraton
Sheath:	Plastic
Alternativen:	Warthog Folder
Manufactured in:	USA
Supply Number:	none (privately procured)

This soldier is a member of the 525th Battlefield Surveillance Brigade. The brigade's mission is to gather information on the enemy. On his vest the soldier is carrying a Kershaw Boot Knife.

KERSHAW MILITARY BOOT KNIFE

Military Knife with German Roots for Proponents of Traditional Forms

The Kershaw K4351 Military Boot Knife with sheath.

The Military Boot Knife by Kershaw can be carried in a variety of ways: with Velcro straps on MOLLE loops, under the shoulder with the handle down (with standard shoulder strap), on the calf or on the arm. The knife cannot, however, be clipped to the boot, as its name suggests. The designation "boot knife" is based on the knife's classic shape.

In the 1980s the original knife was made by Anton Wingen/Othello of Solingen. It was also a drop-forged, full-tang knife with a leather sheath. Well-preserved examples of the original knife fetch high prices at international auctions. The Kershaw Military Boot Knife is an exact copy made in China. 1.4116 steel is used for the blade and tang. This steel, containing about 0.5% carbon and 14.5% chromium, is considered very easy to maintain. Its edge-retention qualities are not all that good, however, but acceptable. During manufacture, a hot metal blank is placed in a two-piece hollow form (upper and lower dies) which has the shape of the knife. A forging press closes the die and shapes the blank.

In a fully-integral knife like the Kershaw model, blade, hand guard, handle and pommel consist of a single piece of steel. This design style is extremely robust. Polyoxymethylen (POM), a thermoplastic material, is used for the inlaid scales. This material is very hard and retains its strength to minus 40 degrees. One will look in vain for riffles or finger choils on the Military Boot Knife, and the hand guard is also not particularly prominent. The relatively slim, almost oval-shaped handle does, however, permit quick changes of grip. The sheath is made of plastic.

The spine of the 11.5-centimeter-long blade is not sharpened, having instead a false edge. The blade tip is above the blade centerline and relatively thin. This is no combat knife for heavy tasks. The Military Boot Knife can, however, be a good choice for an emergency or defense knife.

SPECIFICATION

Manufacturer:	Kershaw Knives – Kai USA
Model:	K4351 Military Boot Knife
Blade Length:	115.0 mm
Blade Thickness:	4.0 mm
Blade Steel:	1.4116
Overall Length:	229.0 mm
Weight (with/without sheath):	–/130 g
Handle:	POM
Sheath:	Plastic
Manufactured in:	China
Supply Number:	none (privately procured)

The 39th Infantry "Bowie Brigade" is a brigade of the Arkansas Army National Guard. From March 2004 to April 2005 the unit operated under the command of the 1st Cavalry Division in an area bordered in the north by the city of Baghdad. The brigade's officers from the rank of major and NCOs from the rank of sergeant-major carry as a status symbol a Bowie knife made specially for the brigade. Here Sergeant-Major Easley is seen with his Bowie knife made by Kenny Teague.

LILE BOWIE BRIGADE KNIFE

The Legendary Bowie Knife of the 39th Infantry Brigade from Arkansas

"There's a rumor going around among the Iraqis that the only way to get this badge is to kill someone with a knife", declared Brigadier-General Ronald Chastain, commander of the 39th Infantry Brigade. With a smile, he added, "But the diamond and the Bowie knife in our unit emblem are just symbols of our home, the state of Arkansas." The diamond in the emblem represents the diamonds found naturally in Arkansas, which has the only diamond mine in North America. The shape of the Bowie knife, for which American national hero Jim Bowie became famous, originally came from Arkansas. During the founding period of the United States, the people of Arkansas were known for carrying large hunting knives and using them at every opportunity. The Bowie knife is also a symbol of the infantry and its ability to achieve its objectives through close combat.

When Colonel Ray Camp, then brigade commander, first saw the 39th Infantry Brigade's new emblem in 1967, he became convinced that the brigade's officers should carry Bowie knives as recognition features. After a petition by the brigade, the army officially approved this privilege some time later. In its search for a suitable knife and a knife maker who could produce it in quantity, the brigade found Jimmy Lile in Russellville. James B. Lile was then considered one of the best knife makers in Arkansas, and his reputation extended far beyond the borders of the state.

For the brigade, Lile designed a Bowie knife with a blade of D2 steel. The shape and length of the blade differed from those of the original Bowie to improve its close-combat qualities. Lile made the cross guard of brass. A contract between the brigade and Lile specified that this Bowie could only be sold to officers of the 39th Infantry Brigade and not on the open market. Only officers from the rank of Major and non-commissioned officers from the rank of Sergeant Major are

Knife maker Jimmy Lile of Russellville was given the job of producing the first Bowie Brigade Knives.

The Arkansas Brigade's unit badge features a stylised Bowie knife in its center.

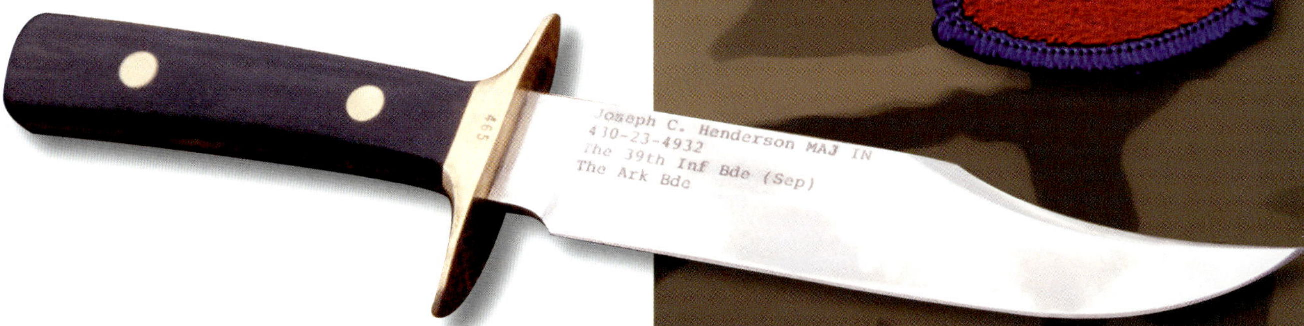

The badge of the 39th Infantry Brigade of the Arkansas Army National Guard features a diamond as the symbol of the state of Arkansas. The only diamond mine in North America is located in Arkansas. It also includes a Bowie knife, for the blade type made famous by Jim Bowie had its origins in the state. Here Major Henderson carries a Bowie knife made for the brigade by Jimmy Lile. The photo was taken at the American base in Taji in December 2004.

permitted to carry a Bowie knife with their uniform. Each knife is engraved with the name of its owner, his Social Security number, his specialty, his rank, the name of the knife maker and the serial number of the knife.

Different materials are used in the handle depending on the rank of the owner. Knives of Sergeant Majors and Command Sergeant Majors are made of cherry wood. Handles for Warrant Officers are made of walnut. For a Major or Lieutenant Colonel the handle is made of black Micarta. A Colonel carries a Bowie with a deer horn handle. The knife carried by the brigade commander is made of ivory. Only knives awarded by order of the brigade commander can be carried by members of the 39th Infantry Brigade. The owner must pay for the knife himself, however.

Jimmy Lile's family continued production of the knife after his death in 1991. In 2003 Lile's former company ceased production, however. The Bowie knives made between Liles' death and the cessation of production had two thin red fiber stripes in the handle. Jimmy Lile and his family made more than 600 Bowie knives for the brigade. At the time, each battalion of the brigade had on average five members entitled to carry the knife. With five battalions in the brigade, adding the members of the brigade headquarters entitled to carry the knife, about 60 Bowie knives are carried by active National Guardsmen.

When Lile ceased production, the officers of the brigade were forced to look for a new knife maker. He was found in the person of Kenny Teague from Mountainburg. Teague learned his craft with the well-known knife maker Bob Dozier. Teague modified Lile' design. His prototype, bearing the serial number 1000 and, like Lile's Bowie knife, made of D2 steel, had a somewhat shorter blade. Teague also modified the handle and made it more ergonomic.

As ivory is now almost unobtainable, Teague gave his knives handles made of linen Micarta and Pakka Wood. The color of the Micarta indicates the rank of the owner: red Micarta is used for non-commissioned officers' knives, black Micarta for those of Warrant Officers, Majors and Lieutenant Colonels. Colonels are issued Bowies with brown Micarta scales, while white Micarta is reserved for Generals. Teague's Bowie knives are delivered with a black leather sheath for dress uniform. The customer can also

SPECIFICATION

Manufacturer:	Lile Knives / Kenny Teague Knives
Model:	39th Infantry Brigade Bowie
Blade Length:	178.0 mm
Blade Thickness:	4.8 mm
Blade Steel:	D2
Overall Length:	308.0 mm
Weight (with/without sheath):	–/–
Handle:	various natural handle materials and Micarta in a variety of colors
Sheath:	Leather and Kydex
Manufactured in:	USA
Supply Number:	none (privately procured)

The rank and name of the owner are engraved in the flat spine.

Kenny Teague's version of the Bowie knife with red Micarta handle scales.

request a Kydex sheath, which is better suited for carriage in regular service and combat. Teague has so far produced about 25 knives for the brigade, with serial numbers starting at 1001.

The following story proves that the Bowie knife is more than a status symbol: in April 2004 a 39th Infantry Brigade convoy was ambushed by Iraqi militants. The lead vehicle, an HMMWV (Hummer) was destroyed by an IED and three members of its crew were severely wounded. The fourth member of the crew, a machine-gunner, was killed by shrapnel from the IED. During recovery of the dead and wounded under enemy fire, it was discovered that the material of the dead man's uniform jacket had become entangled in the rotating machine-gun mount. Major Pelts, part of the crew of another vehicle in the convoy who was taking part in the rescue and recovery action, pulled his Bowie knife and cut the material of the dead man' jacket. His body was then recovered without further problems. Referring to the use of his knife, Major Pelts is later alleged to have said, "You need a good knife, and you must be able to use it quickly."

The Teague model is delivered with a classic leather sheath.

Staff Sergeant Jessen cautiously examines a tin can modified into an IED (Improvised Explosive Device) using his Marto-Phrobis CUK. IEDs are one of the preferred weapons of the resistance fighters and terrorists and are encountered both in Afghanistan and Iraq. Staff Sergeant Jessen is a member of an EOD team of the 748th Ordnance Company of the 184th Ordnance Battalion.

MARTO-PHROBIS C.U.K. COMBAT UTILITY KNIFE

Uncommon But Functional Military Knife

The design of the Marto-Phrobis CUK (Combat Utility Knife) is based on the M9 bayonet developed by Charles A. "Mickey" Finn of Phrobis (the company with the Dolphin logo). In the beginning it was issued exclusively to the U.S. Navy SEALs. With a length of 14 centimeters, the blade is a good five centimeters shorter than that of the M9. The broad, sturdy clip point blade can be used for many purposes. The grind only begins in the lower half of the blade. A broad fuller (forged) on one side of the blade reduces the knife's weight somewhat, without reducing stability. Two-thirds of the blade spine has saw teeth – actually more of a rasp. The blade ends in a round tang. At the end of the Zytel handle a steel pommel is screwed to the tang. All of the CUK's metal parts are black coated. There are two holes in the crossguard to enable the CUK to be used as a spear.

The round handle of fiberglass-reinforced plastic has five deep grooves, and the surface of the handle is highly textured. Although the handle is round, the deep grooves prevent the knife from rolling in the hand. Like the Fairbairn-Sykes Commando Dagger, however, there is the problem of telling which direction the edge is facing during night operations. The CUK was designed primarily as a field knife, and the role of stabbing weapon is a secondary function.

The Zytel sheath represents a marked improvement over the Phrobis M9 bayonet. The clip sheath mount can be rotated 360 degrees, allowing the CUK to be carried with the handle down. The sheath holds the knife very firmly.

The CUK, which is now made under license by the Spanish company Marto, is highly sought after. In America, well-preserved examples fetch prices of around 500 dollars. At such prices the knife is almost too valuable to use.

The fine teeth are best suited for metal or plastic.

Instead of the M9's bayonet fitting, the CUK has a knurled pommel.

SPECIFICATION

Manufacturer:	Marto-Phrobis
Model:	C.U.K.
Blade Length:	139.0 mm
Blade Thickness:	5.5 mm
Blade Steel:	Carbon steel
Overall Length:	260.0 mm
Weight (with/without sheath):	460/300 g
Handle:	Zytel
Sheath:	Zytel
Manufactured in:	Spain
Supply Number:	none (privately procured)

For the paratroopers of D Company, 1st Battalion "Red Falcons", 325th Airborne Infantry Regiment, the transfer back to Iraq at the beginning of 2007 came as a surprise. From June to December 2006 they and American special forces had hunted terrorists from Tikrit in northern Iraq. The unit then returned to America just before Christmas. At the end of December, however, the paratroopers received new marching orders, and once again their destination was Iraq. 1st Lieutenant Defiori is one of D Company's platoon leaders, seen here during a patrol in February 2007. On his hip he is carrying a Mark V by Masters of Defense.

MOD CQD MARK V ADVANCED TACTICAL FIGHTER

Popular Fighting Knife

With the CQD Mark I and II folding knives and the fixed versions CQD Mark V and VI, knife designer and tactical instructor Duane P. Dieter has gained a large following. CQD stands for Close Quarters Defense. The knife was made by the American manufacturer Masters of Defense, which several years ago was bought up by Blackhawk, a maker of tactical equipment. The only model still being made by Blackhawk is the Mark I.

The Mark V was clearly conceived as a weapon, but as dangerous as it looks, it gives the user many possible ways of dealing with an attack. The 17-centimeter-long blade is made of A2 tool steel with a potential Rockwell hardness of 56 to 57. This makes the 5.7-millimeter-thick blade even more break resistant. The main edge of the modified spear point blade has a slightly recurved shape, and the first quarter of the edge is serrated. From the tip, two-thirds of the blade spine has a second edge, however this is not so finely ground as the main edge. The blade has a boron-carbide coating to eliminate reflections and prevent rust. The ceramic-like coating is very hard and reduces friction while cutting.

The side surfaces of the handle have the so-called "Dieter Side Hilt", which also provides additional support for the thumb. The extremely-shaped hilt not only looks very unusual, it also fills several purposes. The "Striker Hilt" prevents the hand from slipping and protects it against contact with opposing weapons. At the same time, one can exert pressure with it and use it to increase striking force. To spare the blade tip, the Striker Hilt can be used for prying, boring and scraping.

The shape of the handle, which is made of fiberglass-reinforced Nylon, is designed to provide maximum safety. Numerous grooves, riffles and surface texture, finger choils and the large pommel serve this purpose. The pommel of hardened steel can be used as a striking aid.

The Striker Hilt can be used for many purposes.

This version of the fiberglass-reinforced plastic sheath is equipped with a Tek-Lok.

SPECIFICATION

Manufacturer:	Masters of Defense
Model:	Mark V ATAC
Blade Length:	173 mm
Blade Thickness:	5.7 mm
Blade Steel:	A2
Overall Length:	323 mm
Weight (with/without sheath):	560/440 g
Handle:	GFK/Santoprene
Sheath:	Prylon Plastic
Manufactured in:	USA
Supply Number:	none (privately procured)

NATO's Kosovo Force was first deployed in 1999. KFOR consisted of about 55,000 troops from more than 25 countries. There were troops from almost every NATO country and many nations of the Partnership for Peace (PIP) programs and various other countries. The Czech Republic also took part with an operational contingent. This Czech soldier was photographed in Kosovo at the end of 1999. His equipment includes an Uton combat knife.

MIKOV UTON

Czech Combat Knife with Well Thought-Out Details

Metal guard effectively prevents the knife from piercing the sheath.

Unlike other Eastern Block states, the former CSSR always displayed a certain degree of independence in its weapons procurement. While most states simply copied the weapons of the big brother USSR, the Czech weapons industry always developed its own versions. This is very well illustrated by its assault rifles and bayonets. The Czech KFOR soldier in the photo on the left carries on his equipment an Uton combat knife made by the Uton company. Today Mikov still makes knives as well as office equipment.

Weighing 350 grams with a 14-centimeter blade, the knife is well-balanced and fits well in the hand. 420 steel is used for the blade. This steel is often used as the basic material for high-quality knives. It is largely similar to the 1.4034 steel used by the Solingen knife industry and contains between 0.4 and 0.5 percent carbon. The most significant alloying component, between 12 and 14 percent, is chromium, which significantly enhances the steel's resistance to corrosion. The Uton model is available in two finishes, satin and black-coated. The unspecified black coating provides additional protection against corrosion and also eliminates reflection.

A saw and a file are included as standard equipment.

One of the most outstanding qualities of the Uton is undoubtedly the ability to insert a file or wood saw into the end of the handle. The synthetic rubber handle contains a special recess into which the tools delivered with the knife can be inserted. The whole thing is held in place by a bolt, which is attached to the knife by a cord.

The accessories are stored in a front pocket when not required. This is present both on the leather and Cordura sheaths. Since joining NATO, the Czech Republic has taken an increasing part in foreign operations, consequently Mikov has also developed a version of the Uton in desert camouflage.

SPECIFICATION

Manufacturer:	Mikov
Model:	Uton
Blade Length:	140.0 mm
Blade Thickness:	approx. 3.0 mm
Blade Steel:	AISI 420
Overall Length:	250.0 mm
Weight (with/without sheath):	350/– g
Handle:	Synthetic Material
Sheath:	Leather or Cordura
Manufactured in:	Czech Republic
Supply Number:	not known

The external appearance of American soldiers has altered completely during operations in Iraq. The new Army Combat Uniform (ACU) was introduced in February 2005. For helicopter crews there is a special Nomex version, the Army Aircrew Combat Uniform (A2CU). Helicopter crews are also issued the Fire Resistant Environmental Ensemble (FREE), a multi-layer clothing system of heat- and fire-resistant fabrics. This loadmaster of a UH-60L Blackhawk helicopter was photographed in Baghdad in 2008. He is wearing the FREE system jacket, pants and insulated under-jacket. His equipment also includes an Ontario ASEK.

ONTARIO A.S.E.K. SURVIVAL KNIFE SYSTEM

System in Operation

Several years ago the American military began searching for a new aircrew survival egress knife. Among the numerous companies that responded to the request for submissions was the Ontario Knife Company, which entered the race with the Survival Knife System. The word system was used in the designation because the knife not only comes with a sheath that allows it to be worn on a pilot's vest, Load Bearing Equipment, belt or strapped to the calf. It is made of heavy nonflammable plastic and two straps hold the knife firmly in place. The system also includes a rope and belt cutter with screwdriver function which fits perfectly in the left leg pocket of flight suits. It is very useful in rescue operations, when it is vital to quickly free an injured man from his clothing or harness. For improved grip, the belt cutter has three deep finger choils. At the top of the handle there is a sharpener, with which the knife can be kept sharp during operations in the field.

One feature of the knife that is immediately noticeable is the relative lengths of the handle and blade: 13.3 centimeters long, the handle, which is molded directly onto the tang, is a good six millimeters longer than the blade. The tang extends the full length of the handle. In the pommel there is a steel insert (with glass/Plexiglas breaker). Three holes (two in the hand guard, one in the pommel) enable the ASEK to be attached to a shaft.

The blade, which is 4.7 millimeters thick, is made of 1095 carbon steel, which is zinc phosphate coated to eliminate rust and reflections. The blade has a Rockwell hardness of 50 to 54. While this does not enhance edge retention, it makes the fine-grained blade virtually unbreakable. The blade spine is equipped with a regular saw, while the edge has a serrated

The ASEK had to undergo a series of tests before the U.S. military finally gave the OK.

The belt cutter is part of the ASEK system.

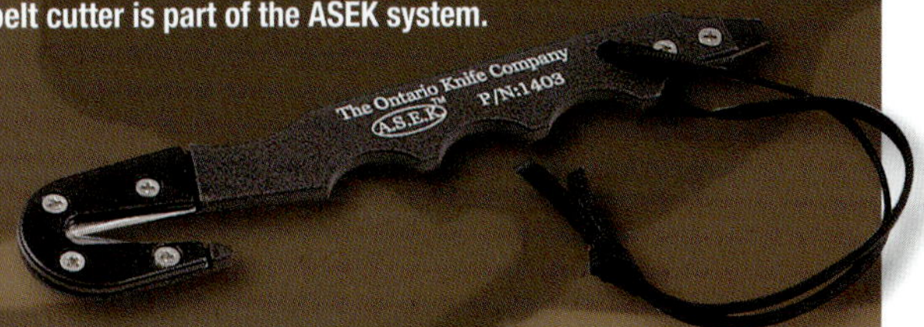

SPECIFICATION

Manufacturer:	Ontario Knife Company
Model:	A.S.E.K.
Blade Length:	127.0 mm
Blade Thickness:	4.7 mm
Blade Steel:	1095
Overall Length:	260.0 mm
Weight (with/without sheath):	–/280 g
Handle:	Synthetic Material/Steel
Sheath:	Plastic/Cordura
Manufactured in:	USA
Supply Number:	NSN 1095-01-518-6832

The 101st Airborne Division (Air Assault) "Screaming Eagles" is one of the U.S. Army's elite units. At the end of 2003 this specialized air assault division was deployed to northern Iraq, the scene of bitter fighting with resistance groups loyal to Saddam Hussein. This machine-gunner was attached to the 1st Battalion, 187th Infantry Regiment "Rakkasans" of the division's 3rd Brigade Combat Team, which was deployed in Tall Afar. On his Interceptor Body Armor combat vest he is carrying an Ontario 498 combat knife.

ONTARIO 498 MARINE COMBAT KNIFE

The Ontario Version of the Most Famous American Combat Knife

Unlike the Ka-Bar Marine Combat Knife, the Ontario 498's leather sheath is black.

During the Second World War Ontario took over the design of the USMC combat knife made by Ka-Bar. The requirement for USMC knives had become so great that the American government contracted several companies, including Camillus, Case and Ontario, to produce this model. Ontario is in fact the official supplier of the U.S. Army.

The Ontario version of this classic knife is equipped with a traditional handle fabricated from leather discs, specially treated to resist the effects of weather – thus the darker color compared to the Ka-Bar model. The 17.8-centimeter blade is made of 1095 carbon steel with a Rockwell hardness of about 57. The non-rustproof but very fine-grained steel is capable of being hardened further (up to 61 HRC), which would increase its edge-retention characteristics. This is not done, however, in order to make the steel somewhat more elastic and easy to care for. With a Rockwell hardness of 57 it is easier to sharpen. For corrosion protection the blade and pommel cap are given a black zinc phosphate coating.

The blade's flat grind only begins in the lower half of the blade, much like Scandinavian knives. To that point the blade is a good four millimeters thick. The knife can thus be used for prying with no ill effects. Only the beak-shaped point is thin, because of the false edge on the spine, and not so sturdy. On the other hand the tip pierces more easily.

The tang extends through the entire handle, with a pommel cap riveted to the end. This pommel cap of tempered steel is suitable for driving tent stakes or as a makeshift hammer. The relatively simple leather sheath has seven rivets and stitching. A handle strap holds the knife in place.

Any knife that has been in service unchanged for more than six decades has undoubtedly proved itself. The low price (about 40 USD on the civilian market) also makes the knife attractive.

SPECIFICATION

Manufacturer:	Ontario Knife Company
Model:	498 Marine Combat Knife
Blade Length:	178.0 mm
Blade Thickness:	4.2 mm
Blade Steel:	1095
Overall Length:	302.0 mm
Weight (with/without sheath):	420/300 g
Handle:	Leather
Sheath:	Leather
Manufactured in:	USA
Supply Number:	NSN 1095-00-392-4102

February 2007: the soldiers of D Company, 1st Battalion "Red Falcons", 325th Airborne Infantry Regiment patrol the streets of Kadamiyah north of Baghdad. Sergeant Willner is one of the unit's paratroopers. His equipment includes an M4A1 carbine and an Ontario 499 knife.

ONTARIO 499 SURVIVAL KNIFE

Good, Simple Survival Knife with an Interesting History

The U.S. Air Force Model 499 Survival Knife has been in service with the American military for several decades and is still extremely popular with members of the army and air force. This is probably because the 499 is relatively inexpensive (about 40 dollars) and exudes a confidence-inspiring toughness. During the Vietnam War the Survival Knife was one of the most commonly carried combat knives.

Overall length is just over 24 centimeters, with the clip point blade accounting for 12.7 centimeters. American 1095 steel is used for the blade and full tang, which have a zinc phosphate coating. Blade thickness is about five millimeters. While the duckbill-shaped area of the spine is ground sharp, the rest of the spine is coarsely serrated, which is particularly useful for cutting aircraft aluminum.

The cross guard has two holes for attachment of the knife to a shaft. Leather discs are used for the handle. A heavy steel plate is riveted to the butt end of the handle, enabling the knife to be used as a makeshift hammer. The sheath is not particularly well-designed: the only thing holding the knife in place is a handle strap with snap. If the sheath is attached to clothing or equipment with the handle down, the knife usually slides out slightly, as in our photo on the left. On the reverse side of the sheath there is a belt loop which comes with a small sharpening stone.

By now many members of the USASFC (United States Army Special Forces Command – Airborne) carry the more modern ASEK, also made by Ontario. From today's standpoint the new model is certainly the better knife, although the Model 499 still has its fans.

The author learns firsthand that the M499 is also carried by pilots of the USMC.

The old-fashioned sheath of the Ontario 499 Survival Knife.

SPECIFICATION

Manufacturer:	Ontario Knife Company
Model:	Mod. 499 Survival Knife
Blade Length:	127.0 mm
Blade Thickness:	4.8 mm
Blade Steel:	1095
Overall Length:	241.0 mm
Weight (with/without sheath):	–/– g
Handle:	Leather/Steel
Sheath:	Leather/Metal
Alternatives:	modern ASEK
Manufactured in:	USA
Supply Number:	NSN 7340-00-098-4327

A member of the U.S. Army keeps watch for enemy movements in his sector. Clearly visible at the bottom left of the photo is his Ontario Mark 3 combat knife, which he has attached to his combat vest with a piece of cord.

ONTARIO MARK 3 NAVY

This Classic Is Even Resistant to Salt Airanhaben.

The Ontario Mark 3 Navy is the official knife of the U.S. Navy and is also used by special units. As the knife is constantly exposed to moisture and salt air in military service, the blade and tang of the Mark 3 Navy are made of 440A steel instead of 1095 carbon steel. While 1095 is somewhat stronger and fine-grained, when it comes to corrosion resistance it cannot compare to the simple 440A steel. 440A contains 0.6 to 0.75 percent carbon and the most important alloying component is 17 percent chromium. This chromium content is responsible for making the steel very rust-resistant. As well, when added to steel, chromium forms chromium carbide, which makes the steel more durable. The blade is hardened to about 56 Rockwells and is given a black coating to enhance resistance to corrosion.

The flat-ground edge begins tapering toward the tip two-thirds of the way along the blade. As a result, the blade retains its maximum thickness of 4.2 millimeters over most of its length, which makes it very rugged. This robust construction comes at some price in cutting efficiency, but this blade geometry is entirely satisfactory for most cutting tasks (rope, plastic, supply rations) and also for self-defense in an emergency. The blade spine is ground to improve the duckbill-shaped tip's piercing ability. The saw tooth edge on the blade spine is suitable for cutting fibrous material and even wood.

The tang extends to the end of the handle, where the scales are screwed in place. Tough synthetic material is used for the handle, which is cruciform-shaped for improved grip. A cross guard prevents the hand from sliding onto the blade, while the wide pommel keeps the knife from slipping, while chopping for example. A lanyard can be attached using a hole in the pommel.

The simple sheath is also made of tough synthetic material and has a nylon retaining strap. The sheath can be attached to the American web belt using the brass clip. The Mark 3 Navy is a simple but thoroughly usable and tough knife, which has proved itself, especially in wet regions.

The Mark 3's fine saw tooth edge can be seen in this photo.

The massive pommel can be used for hammering.

SPECIFICATION

Manufacturer:	Ontario Knife Company
Model:	Mark 3 Navy
Blade Length:	153.0 mm
Blade Thickness:	4.0 mm
Blade Steel:	440A
Overall Length:	277.0 mm
Weight (with/without sheath):	420/280 g
Handle:	Synthetic Material
Sheath:	Synthetic Material
Manufactured in:	USA
Supply Number:	NSN 1095-00-391-1056

This Lance Corporal is part of the crew of an M777A2 155-mm field howitzer of the 1st Battery of the 3rd Battalion, 11th Marine Regiment. The unit was stationed in Iraq at the end of 2007-beginning of 2008. A gun crew consists of eight marines. This gunner is wearing an Ontario SP6 on his ballistic vest.

ONTARIO SPEC PLUS SP6 FIGHTER

Solid Tool from a Large Family.

The SP6, part of the Spec Plus series by Ontario, was essentially designed to be simple and reliable. The knife has an overall length of 33.3 centimeters, of which eight inches (20.3 cm) is taken up by the blade. The knife has a full tang, consequently there is no need to worry about the knife breaking at the junction of handle and blade during heavy work..

With a maximum blade thickness of 4.7 millimeters, the knife is also very sturdy. The 1095 steel used in the blade is standard in many military knives, especially with powder coating, which the SP6 Fighter has. This steel has a good percentage of carbon, with smaller amounts of sulfur (0.5%), phosphorous (0.4%) and manganese (0.6-0.9%). It can be hardened to 58 Rockwells, takes a fine edge, is tough and also has good edge retention characteristics.

The Kraton handle is molded directly onto the tang. The rubberized outer surface ensures a firm grip, even in cold or wet conditions. The oval handle has five grooves to prevent slippage, and at its forward end a cross guard prevents the hand from sliding onto the blade. The curved pommel keeps the knife from slipping from the hand.

Because of the false edge on the blade spine, which extends over most of the blade, the tip is extremely sharp. Nevertheless, the blade's typical Fighter shape is not just suitable for close-quarters fighting, as its name suggests. The knife is also capable of performing daily tasks. The blade is hollow ground on both sides, with the grind set low on the blade.

The sheath is made of Cordura and leather. Two straps with snaps secure the knife in the sheath. The SP6 can be bought for about 60 dollars, a small price for such a reliable knife. Ontario's Spec Plus series comprises more than sixty models, including heavy rescue knives and hatchets. With more than 100 years of history, Ontario is one of the most experienced knife makers in the USA. The company also makes old-fashioned machetes and medical instruments.

The knife is least intrusive when carried on the vest.

The SP6's sheath is made of Cordura and leather.

SPECIFICATION

Manufacturer:	Ontario Knife Company
Model:	SP6 Fighter
Blade Length:	203.0 mm
Blade Thickness:	4.7 mm
Blade Steel:	1095
Overall Length:	333.0 mm
Weight (with/without sheath):	–/– g
Handle:	Kraton
Sheath:	Cordura/Leather
Manufactured in:	USA
Supply Number:	none (privately procured)

In December 2007 the 3rd Platoon, I Company, 3rd Battalion, 23rd Marine Regiment was responsible for security in the city of Bahni Dahir on the Euphrates. The 3rd Battalion, 23rd Marine Regiment is a reserve unit. Its personnel are mostly civilians from various professions. Corporal Sexson was attending college when he was called up to active service. On his right side he is carrying an Ontario Spec Plus SP13

ONTARIO SPEC PLUS SP13 TANTO

Powerful Fighting Knife with Modern Tanto Blade

Founded in 1899, the Ontario Knife Company began producing knives for the U.S. Army during the Second World War and today is one of its most important suppliers. In fact, Ontario supplies knives and bayonets to all four branches of the armed forces (army, air force, marines and navy).

The various models of the Spec Plus series are frequently seen. The SP 13 Tanto Eight has an eight inch (20.3 cm) blade. Along with the two Bowies and the machetes from the same series (each with ten-inch blades), the SP 13 Tanto is one of the largest and heaviest Spec Plus models. Except for its blade, it is similar to the SP6 Fighter.

All Spec Plus models share common features: the wide, thick blade is made of non-rustproof, fine-grained 1095 steel with a "baked-on" epoxy powder coating. A full tang. Kraton or tough Grivory plastic resin handles. Sheaths are made of Cordura and leather.

The SP 13 Tanto's blade is just under 4.8 millimeters thick and its shape has been strongly influenced by Japanese fighting knives. The very slightly curved edge is not rounded toward the tip (as on Japanese Tantos), instead the main and tip edges form an angle (about 130 degrees). The double-sided flat grind begins at about the center of the blade. This makes the knife very strong, as the blade retains its full thickness to that point. With no false edge, the tip is also so stable that it can withstand bending loads well.

A cross guard prevents slippage in the direction of the blade. The prominent pommel ensures a firm grip while chopping. Five deep grooves and the oval handle profile make the Tanto rest securely in the hand. The soldier in the photo on the left has also wrapped a band around the handle of his knife. The sand-colored Cordura sheath, purchased separately from Spec-Ops, has eight grommets and a loop at the tip, so that the knife can also be attached to a tactical vest.

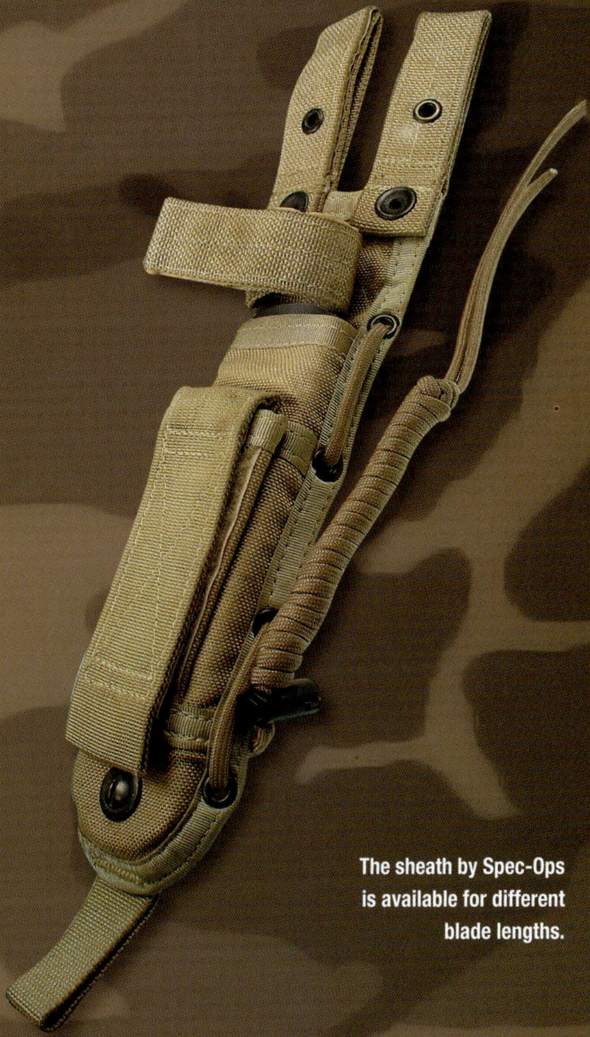

The sheath by Spec-Ops is available for different blade lengths.

SPECIFICATION

Manufacturer:	Ontario Knife Company
Model:	SP13 Tanto
Blade Length:	203.0 mm
Blade Thickness:	4.7 mm
Blade Steel:	1095
Overall Length:	333.0 mm
Weight (with/without sheath):	ca. 400/260 g
Handle:	Kraton
Sheath:	Cordura/Kydex
Manufactured in:	USA
Supply Number:	none (privately procured)

Lance Corporal Johnsen is a machine-gunner in the 3rd Battalion, 23rd Marine Regiment. In December 2007 he and his comrades operated in the Haditha area in western Iraq. He is armed with a 5.56-mm M249 SAW. In addition to ammunition pouches for his light machine-gun, on his Modular Tactical Vest (MTV) he carries a MINI-Flair pen gun and an Ontario Spec Plus SP 25.

ONTARIO SPEC PLUS SP25 USN-2

The Sturdy and Proven Military Knife That Gives Aircraft Aluminum No Chance

The U.S. Marine in the photo on the left is carrying a classic: the SP25 USN-2 has existed in very similar form since the 1950s. In the past, however, the handle of this military knife was made of leather discs, and the sheath was also leather. Since then a change has been made to Cordura (with plastic liner), and on the front there is now a pocket containing a sharpening stone. The sheath in the photo is not an original, rather a cheap replacement from another knife.

The standard sharpening stone is quite capable of maintaining the edge, as the 1095 carbon steel is easily sharpened. The blade spine is 0.188 inches (about 4.7 mm) thick and this is maintained over large parts of the blade. The top third of the broad, epoxy powder-coated blade has a fuller, which helps reduce weight without affecting strength. There is a fine saw tooth edge covering about three-quarters of the spine. This is not designed for cutting wood, however. Instead, the purpose of the saw teeth is to cut aircraft aluminum, to rescue personnel from downed aircraft or to cut pieces of wreckage to use in signaling.

The full tang extends through the Kraton handle, with a steel cap screwed on the end – a proven design used in many military knives. The sturdy cap can be used for hammering. It also has a lanyard hole. The latter is too small for 550 Paracord, however, consequently the SP 25 is often equipped with a leather lanyard.

The knife fits the hand very well and is extremely wieldy due to its short blade (12.7 centimeters) and relatively light weight. It can be purchased in the USA for about 50 dollars. The SP 25 USN-2 has now found a worthy successor in the ASEK pilot's knife.

Stylistically, the SP 25 sheath is based on the leather sheath of the Model 499.

SPECIFICATION

Manufacturer:	Ontario Knife Company
Model:	SP25 USN-2
Blade Length:	127.0 mm
Blade Thickness:	4.8 mm
Blade Steel:	1095
Overall Length:	241.0 mm
Weight (with/without sheath):	ca. 400/260 g
Handle:	Kraton
Sheath:	Cordura/leather
Alternatives:	ASEK pilot's knife
Manufactured in:	USA
Supply Number:	none (privately procured)

The UH-60L Blackhawk is a multipurpose helicopter with a crew of four: pilot, copilot and two loadmasters. Seen here is a loadmaster of the 1st Cavalry Division, photographed at Taji airfield in Iraq in December 2004. He is carrying an Ontario Spec Plus SP 26.

ONTARIO SPEC PLUS SP26 USN-3

Pilot's Survival Knife Issued as Basic Equipment

Ontario's Spec Plus line includes more than 20 models, all outdoors and military knives, which combine proven design elements with modern handle materials and sheath systems. This model's Bowie blade shape is typical for Ontario. The strong, broad blades are given a thick epoxy powder coating, Kraton is used for the handle, and the Cordura sheaths come with long belt loops and a leg strap.

The USN-3 is largely similar to the Air Force Survival Knife 499. The blade is made of non rustproof 1095 steel and measures 12.7 centimeters. Instead of a phosphate coating, it has a much more durable black epoxy powder coating. There is a saw tooth edge on part of the spine, followed by a cord cutter. The tang extends to the end of the Kraton-coated plastic handle. The crossguard is an integral part of the handle. The handle ends in a proven six-cornered steel pommel, which is capable of breaking Plexiglas.

The knife is usually delivered with a modern steel-reinforced Cordura sheath with sharpening stone. The soldier on the left, however, is carrying the SP26 in a leather sheath, which comes with the survival vest issued to American pilots. That is no problem, as the leather sheath was designed to accommodate a variety of knife types. As a rule, a red cord is also attached to the knife too prevent loss. If the pilot drops the knife while struggling to survive, the cord can be used to retrieve it. Sometimes the simplest things are the most practical, without high-tech features.

Like so many knives carried in the field, the SP 26 is nothing spectacular. Its durable coating makes it easy to care for, however, and the powerful blade of low-alloy carbon steel is virtually indestructible and easy to sharpen. The handle provides a firm grip, and the knife can be used not just for cutting, but for hammering and sawing as well. That's enough!

The SP 26 uses the same sheath as the SP 25 model. Modell SP25.

SPECIFICATION

Manufacturer:	Ontario Knife Company
Model:	SP26 SPEC PLUS USN-3
Blade Length:	127.0 mm
Blade Thickness:	4.8 mm
Blade Steel:	1095
Overall Length:	241.0 mm
Weight (with/without sheath):	ca. –/– g
Handle:	Kraton
Sheath:	Cordura/Leather
Alternativen:	A.S.E.K.
Manufactured in:	USA
Supply Number:	none (privately procured)

Unfortunately, we have no details concerning this photo. It was made available by Gary Randall and according to him depicts a U.S. Navy SEAL in Iraq. He is armed with an M21 sniper rifle with a caliber of .308 Winchester. On the rifle is mounted a Leupold scope. Attached to his tactical vest is a Randall Model 14 in a Cordura sheath by the company Spec-Ops with the grip facing down. *Photo: Randall Knives*

RANDALL MOD. 14 ATTACK

A Seldom Seen but First Class Combat Knife

Although the combat knife illustrated here costs about 500 American dollars, it is one of the best known and most sought after knives by a knife maker known worldwide: it is the Model 14 Attack, which has been made since the 1960s by Randall Made Knives of Orlando, Florida. At present the backlog of orders is such that there is a waiting period of up to six years after ordering a Model 14.

This success story began with a first knife, ground from a file and based heavily on the example of Bill Scagel, the grandfather of all knife makers. Then Bo Randall made a knife for a combat soldier, which brought him more orders. Years later Bo Randall even made knives for the first manned space missions. Bo Randall died in 1989 and his son Gary took over the business.

The Model 14 is a nearly indestructible all-round knife designed as a military and survival knife. The forged blade is made of either O1 carbon steel or stainless steels such as 440C, ATS-34 or 154-CM. The O1 steel has a very fine structure, has great tensile strength and is easy to sharpen. The type of steel used is stamped on the blade: an S indicates stainless steel.

The flat grind begins at about half the blade width. A long false edge on the spine characterizes the fighter blade and increases ease of penetration. However, because of the generous spine thickness of about six millimeters, the tip is very sturdy and able to withstand lateral loads.

The broad cross guard is made of brass and is soldered to the handle to keep out moisture and dirt. The two centimeter-wide tang extends into the Micarta handle, where it is held in place with epoxy glue. The standard handle has four finger choils. Other handle shapes, such as "Single Finger Grip" and "Border Patrol" are available on request. At the end of the handle there is a brass grommet for attaching a lanyard.

The knife comes with a high-quality leather sheath with a small front pocket containing a sharpening stone. A strap running diagonally across the crossguard with snap holds the knife in the sheath.

Gary Randall, owner of the legendary knife-making company from Florida.

The company logo, stamped deep into the steel.

SPECIFICATION

Manufacturer:	Randall Made Knives
Model:	Mod. 14 Attack
Blade Length:	190.0 mm
Blade Thickness:	6.0 mm
Blade Steel:	0-1 or 440C/ATS-34/154-CM
Overall Length:	305.0 mm
Weight (with/without sheath):	ca. 620/400 g
Handle:	Micarta
Sheath:	Leather
Manufactured in:	USA
Supply Number:	none (privately procured)

This marine is part of patrol from G Company "Raiders", 2nd Battalion, 2nd Marine Regiment. On the day of the Iraqi parliamentary elections, 15 December 2005, the unit was operating in Al Karmah, a small city north of Fallujah. The marine's Smith & Wesson knife can be seen on his tactical vest.

SMITH & WESSON H.R.T. COMBAT/SURVIVAL KNIFE

Heavy Camp Knife for Tough Jobs

This soldier has attached his HRT using zip ties.

The fixed Smith & Wesson knife has the surname HRT, which stands for "Hostage Rescue Team". The Hostage Rescue Team is an elite unit of the FBI which specializes in anti-terror and hostage rescue operations. The HRT's motto is "Servare Vitas": To save life.

The Smith & Wesson is made by Taylor Brands (www.taylorbrandsllc.com), whose products are largely produced in Asia. With the Hostage Rescue Team's range of duties in mind, an entire series of models was produced, including this massive knife which is suited to difficult tasks. It stands out from the many S&W models in that it was developed by knife maker Darrel Ralph.

With a blade thickness of five millimeters and a more than 17-centimeter-long recurved blade of 440C steel, the HRT can be used for chopping and prying as well as cutting. This full tang model certainly makes an excellent camp knife and position-making tool. The grind begins near the spine, which ensures good cutting characteristics, and the surfaces have a sandblasted finish to reduce reflection.

Scales of Zytel plastic are screwed to the full tang, and the handle's textured surface ensures a firm grip, even with wet hands. The finger choils, the cross guard (part of the handle) and the pommel provide additional grip sureness – especially when the knife is used for rough work. The handle is 16.5 centimeters long, and at the butt end there is a lanyard hole.

The Zytel sheath holds the massive knife securely and can be turned 360 degrees, allowing the user to select the draw angle that suits his taste. A total of six grommets enable attachment cord to be used. The grommet at the tip of the sheath is designed for a lanyard.

Altogether, this is a solidly-made knife with elegant lines, one that is a real workhorse. Unfortunately the SWHRTDR is no longer being made.

SPECIFICATION

Manufacturer:	Smith & Wesson/Taylor Brands, LCC.
Model:	SWHRTDR
Blade Length:	178.0 mm
Blade Thickness:	5.0 mm
Blade Steel:	440C
Overall Length:	243.0 mm
Weight (with/without sheath):	–/–
Handle:	Zytel
Sheath:	Zytel
Manufactured in:	China
Supply Number:	none (privately procured)

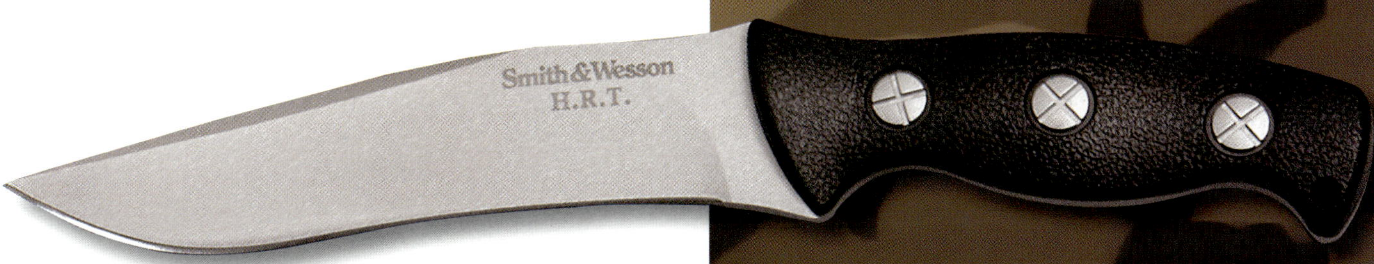

This American soldier is carrying a Smith & Wesson HRT Boot Knife. He is acting as a "breacher" and it is his job to blow open doors using his Mossberg 500 shotgun. The soldier is part of the 4th Stryker Brigade Combat Team of the 2nd Infantry Division and was photographed in Baqubah in early 2008.

SMITH & WESSON H.R.T. BOOT KNIFE

Combat Knife with Japanese Design Elements

The Smith & Wesson Boot Knife from the HRT series is produced in the modern Tanto style. The Smith & Wesson knives made in Asia are relatively inexpensive, and their finish is thoroughly satisfactory.

The main edge of the boot knife is slightly concave, and the spine with its false edge curves slightly downward to the Tanto tip. The grind begins approximately one centimeter below the spine, consequently full thickness is retained over much of the blade, which enhances its strength. The black Teflon coating of the 440C blade helps avoid undesirable light reflection, minimize friction and prevent rust.

The flat, 10-centimeter-long handle is made of aluminum. The surface is textured to provide a firm grip. Optically, the handle is similar to the classic Cold Steel designs, except that they are made of Kraton plastic. A circular cross guard protects the hand from slipping onto the blade. The full tang projects a good centimeter beyond the pommel and forms a lanyard hole.

Black leather is used for the relatively simple sheath, into which the Tanto knife fits up to the crossguard. A strap with snap prevents the knife from being lost. There is a metal pocket clip on the back of the sheath for attaching it to a belt, bootleg or – as in the photo left – to MOLLE loops. Despite this, it is advisable to further secure the knife with zip ties or Paracord.

The Smith & Wesson HRT7T is a reasonably-priced knife made in China, but one that can easily be used in action. Its desirability is enhanced by the use of 440C steel, still one of the best blade steels. If properly hardened, the blade remains sharp for a long time and can also be sharpened in the field without difficulty.

When carried in this manner, the draw angle is not optimal, however the knife is well protected against loss.

A leather sheath similar to the knife in quality is standard equipment.

SPECIFICATION

Manufacturer:	Smith & Wesson/Taylor Brands, LCC.
Model:	SWHRT7T
Blade Length:	121.0 mm
Blade Thickness:	3.8 mm
Blade Steel:	440C
Overall Length:	229.0 mm
Weight (with/without sheath):	–/– g
Handle:	Aluminum
Sheath:	Leather
Manufactured in:	China
Supply Number:	none (privately procured)

Security check at the gate of Camp Warehouse in Kabul in March 2003: a member of AUCON/ISAF, the Austrian contingent of the International Security Assistance Force (ISAF) uses a metal detector to search an Afghani for concealed weapons. Most of the troops of the Austrian ISAF contingent were provided by Jägerbataillon 25. The battalion is the only unit of the Austrian Army trained in airborne landings. This paratrooper is carrying a SEAL 2000 by SOG on his tactical vest.

SOG SEAL 2000

Developed for Maritime Special Operations

Only a few hundred United States Navy SEALs saw action during the Vietnam War, attached to the Military Assistance Command Vietnam – Studies and Observation Group (MACV-SOG), the military high command for Vietnam. They were equipped with a special combat knife that served as the inspiration for the SEAL 2000, which was introduced by the knife and tool maker SOG in the 1990s.

The SEAL 2000 Bowie knife won the American government competition for a new Navy SEAL combat knife and was issued to members of this special unit as standard equipment. SEAL is an abbreviation of the words Sea, Air, Land. And while many manufacturers claim to be the official suppliers of a military special unit – in the case of this model it is true.

Officially the SEAL 2000 is also called the SK2000, the SK standing for Steel Knife. This is because Mission Knives also makes a SEAL knife made of titanium. Although it is a relatively good all-round knife, the SEAL 2000 is seen primarily as a self-defense knife for close combat. The shape of the tip and the blade spine are well suited for controlled stabbing, and penetration force is high. The spine near the tip is not ground to a razor edge, but the user can do this himself.

The knife is delivered with a reinforced nylon sheath, which is not as secure as a hard plastic or Kydex sheath. On the other hand it allows the knife to be drawn or placed back in the sheath silently. Two straps with snaps secure the knife when it is carried handle-down.

Early examples used 440A steel, but this was later replaced by AUS-6. Both are relatively soft but good-natured types of steel, easily sharpened and robust. The blade tends to bend rather than break. However, it is difficult to bend even a simple steel when it is just under 6.4 millimeters thick, as is the thickest part of the SK2000's blade. Maximum blade width is 31.7 millimeters and the weight of the knife without sheath is 354 grams. The flat ground blade has a serrated edge just in front of the handle.

The SEAL 2000 is no longer made, however its replacement, the SEAL Team S37, is quite similar in design. The new knife uses harder AUS-8 steel, however.

The sheath can be attached to equipment already being worn using the Fastex buckle.

When carried behind the ammunition pouches, the SEAL 2000is almost impossible for an opponent to grasp.

SPECIFICATION

Manufacturer:	SOG Specialty Knives & Tools, Inc.
Model:	SEAL 2000
Blade Length:	178.0 mm
Blade Thickness:	6.4 mm
Blade Steel:	AUS-6
Overall Length:	312.0 mm
Weight (with/without sheath):	450/345 g
Handle:	Zytel
Sheath:	Nylon
Options:	SEAL Team S37 successor knife
Manufactured in:	USA
Supply Number:	none (privately procured)

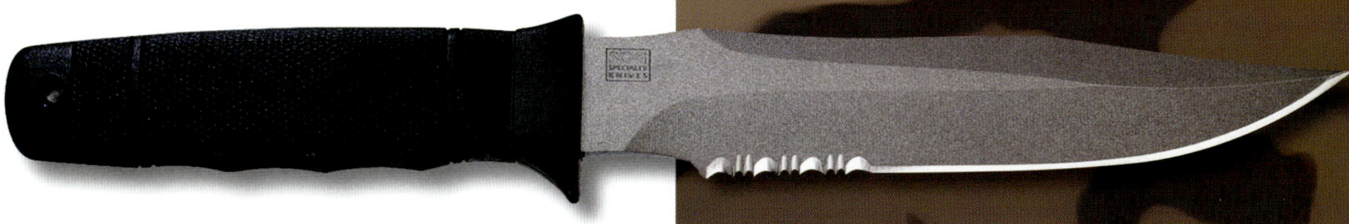

Patrols are part of the daily routine for the U.S. Marines in Iraq. During patrols the Marines regularly visit checkpoints manned by Iraqi security forces. This photo was taken at an Iraqi checkpoint at the bridge over the Euphrates in Bahni Dahir. During a brief rest, a sergeant of the 3rd Platoon, I Company, 3rd Battalion, 23rd Marine Regiment enjoys a cola and a cigarette. On his left thigh he is carrying a SOG Seal Pup.

SOG SEAL PUP

Easily-Carried and Multifunctional Combat Knife

The knife in the photo on the left is the Seal Pup model by the company SOG. The Seal Pup is a lighter version of the SEAL 2000, which was officially introduced for the U.S. Navy SEALs after thorough testing for reliability. This included break tests of the blade and tip as well as examinations of the handle's toughness and the knife's susceptibility to corrosion.

The Kydex sheath in the photo is an optional variant which is equipped with a belt strap with snap in addition to a pocket clip. The sheath holds the knife firmly in place. The Seal Pup comes with a MOLLE-compatible nylon sheath as standard equipment. There is a roomy pocket on the front, which is capable of holding a Multi-Tool, for example. A strap with snap secures the knife.

The unique Bowie shape of the blade is a typical feature of knives of the SEAL series. The AUS-6 steel has a carbon content of about 0.6 percent and 14 percent chromium. It has only moderate edge-retention characteristics (like 440A), but it is easy to sharpen, not particularly prone to breakage and is very rust-resistant. Not a bad choice for a knife that has to be sharpened in the field and is used not just for cutting but also for opening boxes or cans. With its powder-coated blade, the Seal Pup is also frequently used by divers.

Maximum blade thickness is juts over four millimeters. The blade spine is not sharp ground (false edge). One feature of the blade that is often criticized is that, at about 2.5 centimeters, the unsharpened part of the blade in front of the handle (Ricasso) is too long, thus wasting available edge length. While the low grind increases the blade's strength, it also impairs cutting performance. The aggressive serrated edge simplifies the cutting of tough synthetic materials and metal.

The Zytel handle has several finger choils and both sides have a slip-proof textured surface. A lanyard can be attached at the butt end of the handle.

A slot in the sheath serves as a line cutter.

A Kydex sheath is also available for the SOG Seal Pup.

SPECIFICATION

Manufacturer:	SOG Specialty Knives & Tools, LCC.
Model:	Seal Pup M37
Blade Length:	121.0 mm
Blade Thickness:	4.1 mm
Blade Steel:	AUS-6
Overall Length:	229.0 mm
Weight (with/without sheath):	–/153 g
Handle:	Zytel
Sheath:	Nylon
Alternativen:	Seal Pup Elite
Manufactured in:	Japan
Supply Number:	none (privately procured)

Specialist Razo has attached a Seal Pup Elite by SOG to his IBA protective vest with IOTV (Improved Outer Tactical Vest). He was one of approximately 144,000 members of the American armed forces in Iraq in August 2008.

SOG SEAL PUP ELITE

The Third Version of the Famous SEAL Knife

The Seal Pup Elite is one of the most modern knives in the SEAL series. While incorporating many design elements of the other models, in detail it has been completely redesigned. The Bowie shape of the blade is very typical for SOG. Like its older brother, the Seal Pup Elite also has a full tang.

Compared to the Seal Pup, the Elite uses AUS-8 steel (instead of AUS-6), with a Rockwell hardness of about 58. The ratio of blade length to usable edge has been optimized on the Elite. Blade thickness was also increased compared to its smaller predecessor, at 4.7 millimeters. Destroying a well-tempered blade of this strength is almost impossible using normal means. Only at the very tip of the blade is the Elite very finely ground, making breakage possible when used for prying. On the other hand, penetrative force is very high. The entire blade spine has a fine serrated edge, which can be used to cut wood or other even soft metal. Altogether there are four blade variations: uncoated with shiny edge, titanium nitride coated with shiny edge or partial serrated edge, and titanium nitride coated in tiger stripe look with partial serrated edge.

The injection-molded handle is made of fiberglass-reinforced plastic (Zytel) and has a slip-proof textured surface. Four finger choils further improve grip. There is a lanyard hole in the pommel.

There are two different sheaths for the knife, both well thought-out. The nylon sheath in the photo on the left comes standard with the knife. The loop on the back allows it to be attached to belts and MOLLE loops. Inside the sheath is a tight-fitting plastic liner. There is a pocket with snap closure on the front which has space for a multi-tool or small flashlight.

In cooperation with the U.S. Navy SEALs, SOG has also developed a Kydex sheath with a cutout exposing part of the blade. This makes it possible to cut ropes without removing the knife from its sheath.

The knife is also produced with a tiger stripe finish on the blade.

The Cordura sheath offers space for extra equipment such as tools or a sharpener.

SPECIFICATION

Manufacturer:	SOG Specialty Knives & Tools, Inc.
Model:	Seal Pup Elite
Blade Length:	123.0 mm
Blade Thickness:	4.7 mm
Blade Steel:	AUS-8
Overall Length:	241.0 mm
Weight (with/without sheath):	–/150 g
Handle:	Zytel
Sheath:	Nylon
Options:	Uncoated, Coated, Partial serrated edge
Manufactured in:	Japan
Supply Number:	none (privately procured)

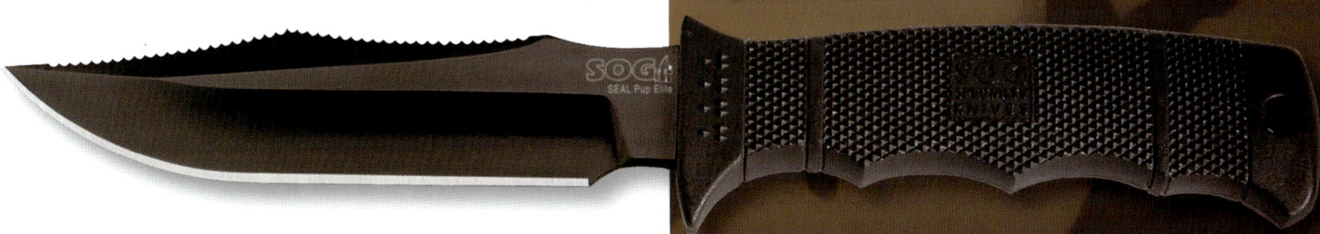

Sergeant Collins is a member of C Company, 2nd Battalion, 4th Infantry Regiment. This unit is part of one of the light infantry battalions of the 10th Mountain Division. In the summer of 2008 the battalion was operating south of Baghdad. The photo was taken during a patrol in the rural outer districts of the city. On his ballistic vest Sergeant Collins is carrying a knife made by United Cutlery.

UNITED CUTLERY ELITE FORCES RECON KNIFE

A Simple, Undemanding Boot Knife

In the photo on the left, the soldier is carrying a rarely-seen knife on the left breast of his tactical vest next to his Benchmade Rescue Hook. The Elite Forces Recon is made by United Cutlery. It is a relatively simple knife made out of a single piece of steel. This integral design results in great strength – especially as maximum blade thickness is a not inconsiderable 4.7 millimeters.

The knife is fabricated from 420J2 steel, which is mainly used as blank material in the manufacture of high-quality knives. This steel is largely similar to the 1.4034 steel often used in Solingen. It contains between 0.4 and 0.5 percent carbon, and the most important alloy metal is chromium (between 12 and 14 percent), which improves resistance to corrosion. An unspecified coating further protects the steel. The center point blade is hollow ground on both sides from the thick central spine. From the tip to the middle of the blade spine there is a false edge. From there the spine – like the handle – is a continuous 4.7 millimeters thick.

The cross guard is wide enough to effectively prevent the hand from sliding onto the blade. This is necessary, as the 11.3-centimeter-long handle is relatively thin and smooth. There is a danger of insufficient grip, especially in wet conditions or when the holding hand is sweaty. The two finger choils improve the grip a little.

The plastic sheath is tight-fitting and holds the knife securely. As well, part of the sheath snaps into the lengthwise cutout in the handle. The sheath comes with a plastic clip to enable it to be attached to boots or LBE loops. With two holes at the bottom end of the sheath, the knife can also be carried with the handle down.

With an official purchase price below 30 dollars, the Elite Forces Recon is economically priced. Despite the relatively unspectacular materials and less than perfect handle, this knife is a useful emergency knife. It is no longer in production, however.

The sheath is made of tough Zytel plastic.

SPECIFICATION

Manufacturer:	United Cutlery Corp.
Model:	Elite Forces Recon Knife
Blade Length:	112.0 mm
Blade Thickness:	4.7 mm
Blade Steel:	420J2
Overall Length:	224.0 mm
Weight (with/without sheath):	–/–
Handle:	420J2
Sheath:	Plastic
Manufactured in:	China
Supply Number:	none (privately procured)

Quite often, Marines or soldiers of the U.S. Army carry two or more knives. Almost every one has a folding knife somewhere. Many soldiers also carry a multi-tool. The third is usually a combat knife, worn on the outside of the combat vest. In addition to his Ontario bayonet, this marine is carrying a Hornet Boot Knife by United Cutlery. The photo was taken in the Fallujah area of Iraq in December 2005.

UNITED CUTLERY HORNET BOOT KNIFE

Simple but Robust Combat Dagger

Both edges of the blade have fine serrated sections.

The tang projects slightly from the handle.

The casual manner in which the dagger is being carried by the soldier in the photo is not recommended, even though the blade is obviously not very sharp. The marine infantryman could easily injure himself, and in close combat an opponent could easily grab the knife.

The Hornet Boot Knife is a classic representative of the 1990s: satin-finished blade, Kraton handle and nylon sheath. Made in various sizes, the dagger was offered by many companies. The knife made in Taiwan was offered for sale in the USA by United Cutlery, among others. The Hornet Boot Knife was already no longer part of the company's line when it became insolvent.

The eleven-centimeter blade is made of 440A steel. The corrosion resistance of this inexpensive steel is quite high, and sharpening is also quite simple. Only its edge retention characteristics are somewhat less than higher quality steels. In keeping with the spirit of the times, the blade was made with a satin finish, probably patterned after the Gerber Mark II, the progenitor of this type of dagger. The same applies to the dual partial serrated edges, intended to give the dagger more "bite". As a rule, the blade geometry of these models precludes a particularly sharp edge.

The handle of Kraton thermoplastic fits comfortably in the hand, but it is quite short in keeping with the boot knife design. Interestingly, the tang extends beyond the pommel and can be used for light hammering. The rubber-like nature of the plastic used in the handle results in a less than stable shape, which deforms under pressure. There is a lanyard hole at the butt end of the handle.

The nylon sheath is of simple design with a plastic liner. A snap fastener is supposed to prevent the Hornet from being lost. Even though the sheath is less than spectacular, the marine in the photo would be recommended to use it for carrying the knife.

SPECIFICATION

Manufacturer:	United Cutlery Corp.
Model:	Hornet Boot Knife
Blade Length:	110.0 mm
Blade Thickness:	5.0 mm
Blade Steel:	440A
Overall Length:	220.0 mm
Weight (with/without sheath):	180/130 g
Handle:	Kraton
Sheath:	Nylon
Manufactured in:	Taiwan
Supply Number:	none (privately procured)

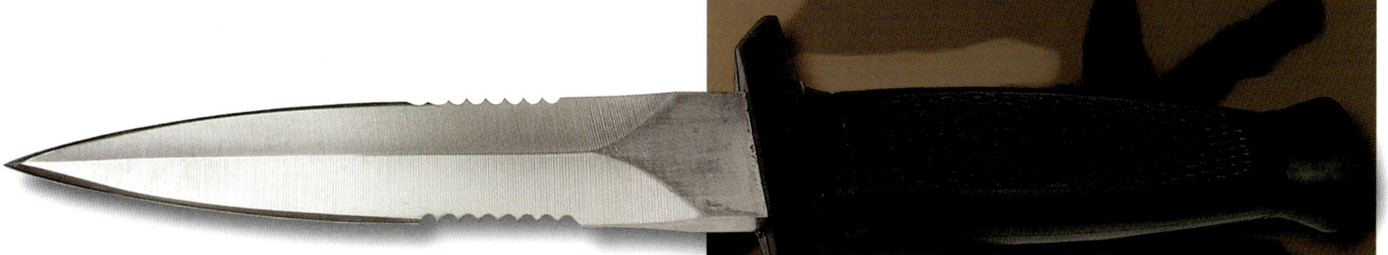

The soldier in this photo is a member of the Swedish ISAF contingent. He is using a PSG90 sniper rifle to cover fellow soldiers as they move forward. The PSG90 is the Swedish model of the well-known .308-caliber Accuracy International sniper rifle. The scope is a Hensoldt model with ten power magnification. He is also armed with a Glock pistol.

Photos: Anton Muhrbeck & Magnus Andersson

VOX KNIVES PRY MATE

Hefty Knife from a Danish Knife Maker

This photo of a Swedish sniper once again demonstrates clearly the wide variety of knives carried by soldiers in war zones. One encounters the entire gamut, everything from simple, lifeless shapes made from a piece of steel to expensive handmade custom knives. An example of a high-end knife is the Vox Pry Mate.

The man behind Vox is Danish knife maker Jesper Voxnæs, who developed a small selection of base models which he customizes as his customers desire. The handle is made of colored G-10 material or Micarta and is equipped with or without fiber liners or titanium hollow rivets. There are six base models, from the "MegaMini" finger knife (overall length 8.9 cm) to the 29-centimeter-long "Voxrold". Jesper previously forged most of his knives. In recent years, however, he has used mainly RWL-34 rapidly solidified powder steel or Damasteel, from which he grinds the blades using his belt sander or by hand.

RWL-34 is distinguished by its carbon content of 1.05 percent and approximately 14 percent chromium. It is the rapidly solidified powder steel counterpart of ATS-34 steel, but because of the way it is manufactured it has a more even and finer grain, which contributes positively to the strength and possible fineness of the edge.

At 6.4 centimeters, the blade is short and very wide and is hollow-ground from its seven-millimeter-thick spine. Seven millimeters is a statement and promises enormous strength. Numerous soldiers contacted Jesper Voxnæs and told him they needed a knife that could be used for prying without damaging the blade. For customers wanting a blade geometry better suited to cutting, there is the "Voxmate". Of similar design, it has a 4.7-millimeter-thick sheepsfoot blade.

There is a finger choil at the junction between the handle and the blade. The tang extends the length of the Micarta handle, retaining its full thickness. Two titanium-lined hollow rivets with a diameter of 13 millimeters join the tang and handle. The rear rivet can accommodate a lanyard. A Kydex sheath with a carbon fiber finish holds the knife securely and can be fastened to a belt.

The Vox Pry Mate's unobtrusive Kydex sheath.

SPECIFICATION

Manufacturer:	Vox Knives
Model:	Pry Mate
Blade Length:	64.0 mm
Blade Thickness:	7.0 mm
Blade Steel:	RWL-34
Overall Length:	127.0 mm
Weight (with/without sheath):	–/–
Handle:	Micarta
Sheath:	Kydex
Options:	variety of handle and blade materials
Manufactured in:	Denmark
Supply Number:	none (privately procured)

Members of the 39th Armoured Engineer Squadron of the 32nd Engineer Regiment overhaul two pontoon bridges over the Schatt al-Arab in the city of Basra. Here the unit's Squadron Sergeant Major works on a bridge ramp. On his back he carries an M-Line KN400 Beta Titanium dive knife by XS Scuba.

XS SCUBA BETA TITANIUM KNIFE

Dive Knife Used by Combat Engineers

This close-up shows the knife attached with zip ties and the red unlocking button on the sheath.

The black Beta titanium knife with the model number KN400-BK (there are also blue and yellow models) is a knife specially developed for divers. It is made by XS Scuba, an American manufacturer of diving equipment. The Beta Titanium is interesting on account of the material used in its blade. Instead of steel, a special titanium alloy is used.

Basically, one can say that steel always has the better edge-retention qualities and is the better blade material, at least when it comes to cutting performance. Special titanium alloys like Beta-Titanium, which contains vanadium and aluminum, are extremely strong and non-brittle, however. As well, titanium is absolutely rustproof. Other interesting qualities are its low weight (about half that of steel for the same volume), and the fact that titanium is non-magnetic, which makes it popular among divers whose job it is to disarm mines.

The Beta titanium alloy used by XS Scuba achieves a Rockwell hardness of 51 to 54. The 12.7-centimeter-long blade is ground sharp on both sides. On the upper edge there is a line cutter and a serrated section. There are two other variants of the knife, one with a blunt screwdriver tip (ground on both sides) and one with a Tanto blade (just one edge).

The handle design is the same on all three models of the M-Line: a slip-proof plastic grip with four finger choils, "tire profile" and a crossguard ensure a firm grip under water. The handle also has a hammer end made of stainless steel.

The double-locking sheath prevents knife loss. It has drain holes and can be fastened to the calf, the upper arm or equipment. While the knife's dagger shape with two edges eliminates it as an all-rounder, it is well-suited for most tasks and as an emergency knife.

A typical plastic dive knife sheath comes as standard with the knife.

SPECIFICATION

Manufacturer:	XS Scuba
Model:	KN400
Blade Length:	127.0 mm
Blade Thickness:	4.0 mm
Blade Steel:	Beta-Titanium
Overall Length:	270.0 mm
Weight (with/without sheath):	–/– g
Handle:	Plastic
Sheath:	Plastic
Alternativen:	Tanto, Blunt Tip
Manufactured in:	Taiwan
Supply Number:	none (privately procured)

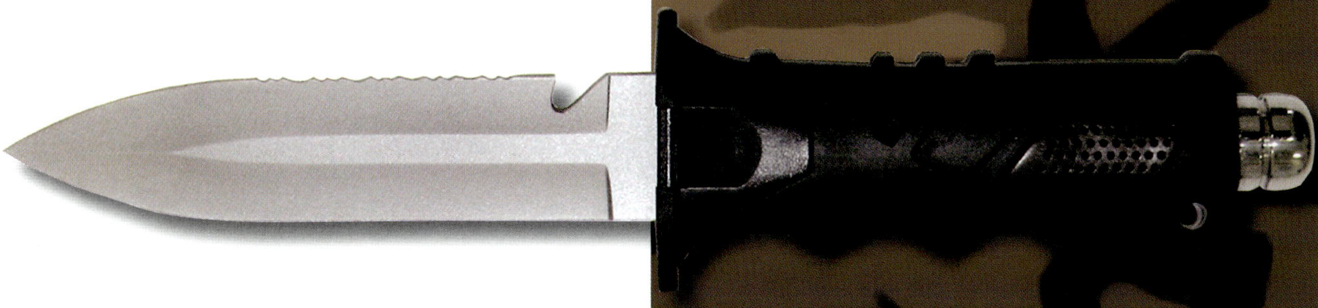

BAYONETS

October 2008, situation briefing. Shortly before a joint patrol with Iraqi troops, Major Allen, commander of the 2/51 Brigade Military Transition Team from the 2nd Battalion, The Royal Anglian Regiment of the British Army, briefs his men on the planned mission into the city of Basra. He is pointing out the march route on a map using his SA80 bayonet. MTTs are platoon-strength training and support teams attached to the battalions, brigades and divisions of the Iraqi armed forces.

Canadian troops of the ISAF contingent in Kandahar, Afghanistan in 2006. The soldier in the foreground is armed with a Colt-Diemaco carbine and a B2000 bayonet. The Canadian flag emblems are IR reflecting, the uniform is Desert CADPAT, and the vest is Standard CADPAT (Canadian Pattern).

Photo: Sören Sünkler

EICKHORN B2000 BAYONET

Multipurpose Bayonet for Modern Infantry Units

The B2000 bayonet by A. Eickhorn GmbH & Co. für Schneidwaren und Waffen KG of Solingen (today The Original Eickhorn Solingen Co. Ltd.) was developed in close cooperation with NATO forces specifically for the extensive scope of duties of modern infantry units. In order to obtain a NATO supply number the B2000 had to undergo rigorous testing. The choice of materials for the bayonet reflects these demands: 55Si7 spring steel with a Rockwell hardness of 51 to 53 is used for the 3.6-millimeter-thick blade. Normally, high-grade knife blades have a hardness of between 56 and 60 HRC. Because of the reduced hardness, the bayonet is not only easy to sharpen but is also highly non-brittle. The 18.4-centimeter blade can be bent 30 degrees in either direction without breaking.

The B2000 is less a cutting tool than a stabbing weapon, upon which one can rely unconditionally in extreme situations. The rationale: better a dull and bent bayonet than a broken one. Nevertheless, the B2000 can also be used for cutting. It is capable of cutting even hardened barbed wire using the built-in wire cutter. An adjustment screw in the cutting plate on the sheath ensures optimum position. The blade and cross guard are Kalgard-coated to prevent rust and eliminate reflections. The B2000 can be mounted on any NATO assault rifle.

The grooved handle is made of fiberglass-reinforced plastic. It is insulated against charges up to 10,000 Volts, is acid and alkaline resistant and can be decontaminated. The handle is thicker at the center and rests firmly and comfortably in the hand. The plastic sheath with belt strap includes a diamond sharpening plate. The B2005CAN has been in service with the Canadian Armed Forces since 2005.

The reverse side of the sheath is MOLLE-compatible.

Wire-cutter system on the sheath.

SPECIFICATION

Manufacturer:	The Original Eickhorn-Solingen Co. Ltd. (formerly: A. Eickhorn GmbH & Co. für Schneidwaren und Waffen KG of Solingen)
Model:	B2000
Blade Length:	184.0 mm
Blade Thickness:	3.6 mm
Blade Steel:	55Si7
Overall Length:	312.0 mm
Weight (with/without sheath):	550/300 g
Handle:	Plastic
Sheath:	Plastic
Manufactured in:	Germany
Supply Number:	NSN 1095-25-149-6922

Sweden also contributed a contingent to KFOR. The Swedes were deployed southeast of Pristina, the capital of Kosovo. This Swedish infantryman is armed with a 5.56 x 45 mm AK5 assault rifle. Beneath the forward shaft of the weapon is a 40-mm grenade launcher. On his vest the soldier is carrying an Eickhorn G3 bayonet.

EICKHORN BAYONET G3

A German Bayonet Developed for the Swedish Military.

The Swedish KFOR soldier in the large photo on the left is wearing a bayonet on his tactical vest which was developed by Eickhorn of Solingen for the Heckler & Koch G3, HK33 and G41 series assault rifles. Interestingly, the soldier is carrying an AK5 assault rifle, a Swedish weapon on which the G3 bayonet will not fit. Years ago the Swedes equipped their armed forces with the new 5.56-mm assault rifle, nevertheless the bayonet remains in use as a pure combat knife.

The G3 was designed as a pure weapon, preferably employed for thrusting. The shape of the blade is similar to that of the American M3 combat knife or the M7 bayonet and is of spear point configuration. 55Si7 spring steel is generally used, with a rough, matte phosphate coating. The bayonet is mounted on the weapon from above, similar to the French FAMAS system. The G3, HK33 and G41 assault rifles have a spring-loaded adapter in the cocking tube, which engages the bayonet's locking pin.

The design of the weapon and of the mounted bayonet makes it necessary to protect the thermoplastic handle against the hot gases exiting the flash suppressor. Without some sort of protection, the high temperatures would very quickly deform the plastic. For this reason, in the area of the handle/guard there is a so-called "fire protection cap" made of metal. On the Swedish version the cap bears the Three Crowns, the national emblem of Sweden.

The plastic sheath holds the bayonet with the aid of a metal spring which engages a slot in the guard. Additional security is provided by a handle strap with snap. Thanks to its U-shaped hook, the sheath can also be attached to the American web belt. Two holes at the bottom of the sheath allow moisture to escape and can also be used to tie the sheath to the leg.

The G3 bayonet is still available from The Original Eickhorn Solingen Co. Ltd., however more and more militaries are equipping their troops with multipurpose bayonets whose use is not restricted that of a weapon.

The G3 bayonet's locking pin.

A slot in the crossguard locks the bayonet in the sheath.

SPECIFICATION

Manufacturer:	The Original Eickhorn Solingen Co. Ltd. (formerly: A. Eickhorn GmbH & Co. für Schneidwaren und Waffen KG of Solingen
Model:	Bayonet G3
Blade Length:	169.0 mm
Blade Thickness:	5.0 mm
Blade Steel:	55Si7
Overall Length:	305.0 mm
Weight (with/without sheath):	510/340 g
Handle:	Plastic
Sheath:	Plastic
Manufactured in:	Germany
Supply Number:	not known

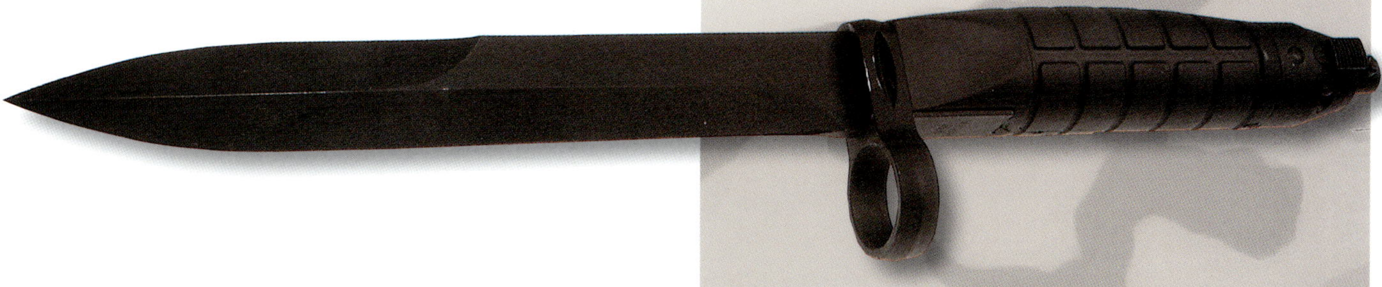

This Captain is an officer of the 2nd Battalion "The Poachers", The Royal Anglian Regiment. This light infantry battalion has been stationed in Germany since the summer of 2007. The photo was taken in 2008 in Al-Qurnah, where a platoon from the battalion was stationed. The captain is carrying the bayonet for his SA80A2 assault rifle without sheath in the elastic loops on the front of his combat vest.

SA80 BAYONET

A Mystery of Procurement Policy

The soldier in the photo on the left is handling his SA80 bayonet rather carelessly. Stuck tip down into the loops on his vest, he is carrying the bayonet without a sheath. This is in keeping with the attitude many soldiers have toward the SA80 bayonet. This attitude can be summer up – with typical British humor – as follows:

The SA80 is a cheap integral bayonet made of the finest metal – melted-down Matchbox cars. Guaranteed to be more breakable than the bones of your ninety-year-old grandmother. Mounted on the muzzle of the assault rifle, the SA80 gets so hot that one can boil a pot of water simply by sticking it in briefly – provided one has an asbestos glove. Its most interesting characteristic is its inability to stay sharp. This is clearly a sign of too-soft steel, but it also has a positive side: the nicks produced by cutting string can easily be pounded out again. The SA80 is one of the modern mysteries of the British Army's procurement policies. One can learn more at www.british-army-military.com.

The SA80 bayonet fits the L85 and L86 rifles produced by the Royal Small Arms Factory and in service with the British Army. Beginning in 2005, the weapons were updated and improved by Heckler & Koch. The bayonet's 17.8-centimeter-long clip point blade has an extremely mild serrated edge over half its length. Above the serrated edge there are fullers on both sides of the blade. The surface is sandblasted matte.

The black painted handle is also made of metal and is permanently attached to the blade. A crossguard protects the hand. Seven grooves around the handle are intended to provide a firm grip. The butt end is hollow, allowing the SA80 to simply be stuck onto the weapon's flash suppressor. The spring-loaded locking button sits under the handle.

The knife is delivered in a nylon case. It contains a sheath body of fiberglass-reinforced plastic with an integral wire cutter that fits into a slot on the blade. The sheath also has a folding saw blade and a diamond sharpening plate.

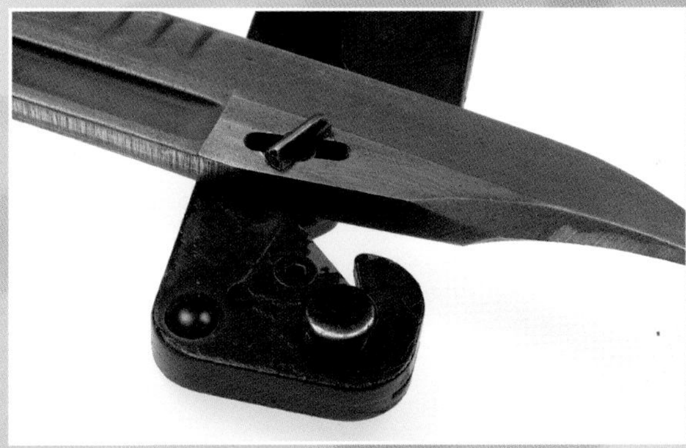

The massive wire cutter.

The integral sharpening surface on the sheath.

SPECIFICATION

Manufacturer:	—
Model:	SA80 Bayonet
Blade Length:	180.0 mm
Blade Thickness:	6.0 mm
Blade Steel:	Rustproof Steel
Overall Length:	284.0 mm
Weight (with/without sheath):	780/420 g
Handle:	Steel, Painted Black
Sheath:	Plastic with Nylon Case
Manufactured in:	Great Britain
Supply Number:	NSN 8465-99-132-1567

From the beginning, French troops have been a part of ISAF. The French contingent grew from 500 troops in 2002 to 3,300 at the beginning of 2009. This member of the crew of an armored car was photographed in the suburbs of Kabul in March 2002. His equipment includes a FAMAS bayonet.

FAMAS BAYONET

The Bayonet of the French Armed Forces

The bayonet is often the subject of heated discussion. To some the bayonet is outmoded in the age of digitalized warfare and is simply no longer contemporary. As well, the bayonet's image does not appeal to many politicians, as it has the reputation of being a brutal weapon. Today the bayonet will not be found among the equipment of Bundeswehr, for example. Other armies see the situation differently, for there is no war that can be fought only at a distance – infantrymen often find themselves in dangerous situations in which man fights against man. And so the French, along with the Americans and British, equip their troops with bayonets.

The FAMAS bayonet (FAMAS is the abbreviation of "Fusil d'Assault Manufacture d'Armes de St. Etienne"), seen here being carried in its sheath handle-up, fits the FAMAS assault rifle, which can also be seen in the photo and which has been in use by the French military since 1977. On account of its compact design and large carrying handle, the recoil-operated weapon has earned the nickname "Le clarion" (bugle) among French troops. Made by GIAT, in modified form the FAMAS assault rifle will remain the primary weapon of French soldiers for some years to come.

The FAMAS bayonet has a plastic handle held in place by two flat-head screws and it attaches to the assault rifle edge-up by means of two mounts. The slim center point blade is only ground on one side. The FAMAS bayonet retains its full thickness over large parts of the blade width. A short part of the blade spine has a false edge. All metal parts have a phosphate coating.

A spring-supported rocker, braced against the flash eliminator, holds the bayonet on the weapon during use. The sheath of fiberglass-reinforced plastic has a friction spring near its mouth, which fixes the bayonet. The sheath attaches to a web belt by means of a U-shaped metal clip. Alternately, the sheath can be attached to any belt with a maximum width of no more than five centimeters.

A friction spring holds the bayonet securely in the sheath.

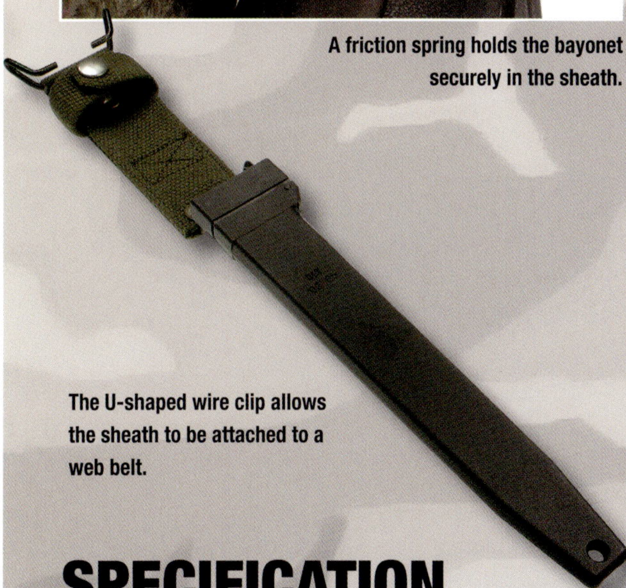

The U-shaped wire clip allows the sheath to be attached to a web belt.

SPECIFICATION

Manufacturer:	Nemex (ex-Giat Industries)
Model:	FAMAS Bayonet
Blade Length:	204.0 mm
Blade Thickness:	4.0 mm
Blade Steel:	Carbon Steel
Overall Length:	327.0 mm
Weight (with/without sheath):	400/300 g
Handle:	Plastic
Sheath:	Plastic
Manufactured in:	France
Supply Number:	Not Known

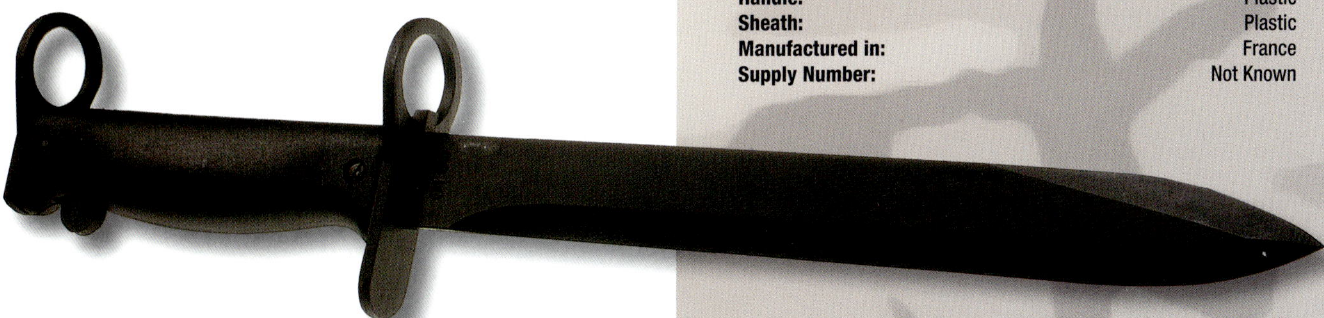

As of April 2004 the Multinational Division Baghdad (MNDB) of the coalition forces in Iraq was led by the 1st Cavalry Division from Fort Hood, Texas. Private First Class Haynes is one of the troopers of D Troop, 9th Cavalry Regiment. In December 2004 the platoon was one of the units providing security in the Green Zone in central Baghdad. He has attached a Kalashnikov bayonet to the shoulder of his tactical vest.

KALASCHNIKOW BAYONET M1974

Successful Multipurpose Bayonet Developed by the Soviets

Even though the assault rifle being carried in this photo is obviously not a Kalashnikov AK-74, the American soldier is carrying a bayonet developed specially for this weapon, used by the Soviet Army, the former East Block States and the present-day Russian Army.

For several years the Bundeswehr was also an "official" user of the Kalashnikov M1974 bayonet. In 1993 paratroopers, together with panzer grenadiers, light infantry and army anti-aircraft troops were equipped with East Block combat knives. With the disbandment of German Democratic Republic's National Volksarmee, the Federal Army also came into possession of countless bayonets produced by the former VEB Fahrzeug und Jagdwaffenwerk Ernst Thälmann in Suhl. The NVA Multipurpose Bayonet M1974 with serrated edge on the spine and wire-cutter on the sheath was introduced into service by the Bundeswehr as the "Combat Knife, Heavy" (supply number 1095-12-700-2611). Even though the M1974 mounting system fit neither the new Bundeswehr web belt nor the G 36 assault rifle introduced in the mid-1990s, the bayonet remained in service for years. Many soldiers used the expedient of adhesive tape to attach the knife to their equipment, for example the shoulder strap of the belt carriage system.

The handle of the AK-74 model series is made of black or brown plastic. The knife can be attached to the wire-cutting attachment on the tip of the sheath by means of the rectangular cutout in the blade. The broad blade is a definite all-rounder. The edge is big-bellied to the tip, and the uniform blade thickness over much of the blade (the grind begins low on the blade, almost like a Scandinavian knife) increases strength and "misuse potential". The serrated spine is useful for at least cutting thinner branches or plastics. The quality of the blade steel is generally considered to be medium to poor. The AK-74 bayonet also has poor edge-retention characteristics, consequently it must be sharpened frequently.

The wire-cutter feature and the serrated section on the blade spine.

Classic metal sheath with leather belt loop and retaining strap.

SPECIFICATION

Manufacturer:	—
Model:	Bayonet M1974
Blade Length:	150.0 mm
Blade Thickness:	3.0 mm
Blade Steel:	Rustproof steel
Overall Length:	273.0 mm
Weight (with/without sheath):	520/310 g
Handle:	Plastic
Sheath:	Plastic
Manufactured in:	Russia
Supply Number:	None (privately procured)

In mid-January 1996, 1,500 Russian troops arrived in Bosnia-Herzegovina. The Russian IFOR contingent was attached to the U.S.-led Multinational Division North (MNDN) and operated in the northeastern part of the war-ravaged country. This Russian paratrooper was photographed at a checkpoint in March 1996. He is armed with an AKS-74 assault rifle with a caliber of 5.45 x 39 with fixed M1989 bayonet.

KALASCHNIKOW BAYONET M1989

The Final State of Development of the Soviet Multipurpose Bayonet

This Russian paratrooper is carrying his M1989 with the handle up.

Unlike the familiar Kalashnikov multipurpose bayonet with its characteristic beak-shaped tip, this bayonet is less commonly seen on Russian soldiers. It first appeared in Afghanistan in the mid-1980s, where it was used by Soviet troops. Bayonets of this type carry the designation M1989 or "6 x 5". Although most of the bayonets associated with the Kalashnikov system were copied by the client states, to date only Bulgaria has decided to reproduce this bayonet.

The M1989's handle is made of black plastic, with a textured surface and three raised beads to improve grip. Like the other Kalashnikov models, this one is also equipped with a wire-cutter.

The narrow, straight blade with the dagger-like tip is primarily intended for use as a weapon, however it can be used for some general tasks. The type of blade chosen for the M1989 has reduced its operational value somewhat compared to the older Kalashnikov bayonets. The serrated spine with 44 teeth is at least usable for cutting thin branches, plastic or sheet metal. It is not known if the quality of the steel has been improved over that of the earlier models. At least the steel used in the M1959 and M1974 models had the reputation of being fair to middling.

The M1989 sheath is similar to that of the M1974 version in design and is a combination of metal and thermoplastic. In order to be able to combine the sheath with the blade to form a wire cutter quickly and simply, the carrying system includes a carbine hook. The ambitious collector of military edged weapons will find it hard to obtain an original M1989, for the Russians have been extremely reluctant to issue export licenses for this bayonet.

SPECIFICATION

Manufacturer:	IschMasch
Model:	M1989 ("6x5")
Blade Length:	160.0 mm
Blade Thickness:	3.5 mm
Blade Steel:	Rustproof steel
Overall Length:	285.0 mm
Weight (with/without sheath):	–/– g
Handle:	Plastic
Sheath:	Plastic
Manufactured in:	Russia
Supply Number:	—

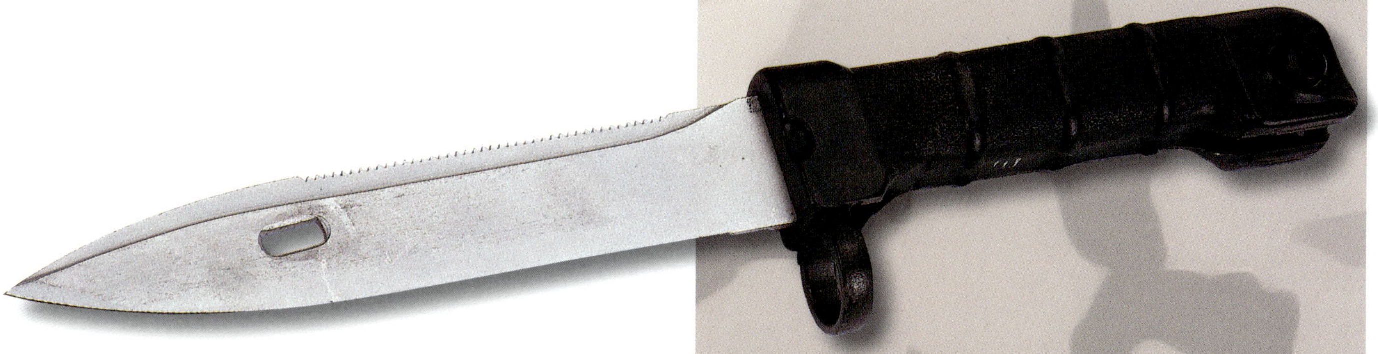

December 2003, Baghdad. Demonstrations followed the first attacks in which Iraqi civilians were killed. American troops and Iraqi police provided security for the demonstrations. One of them was Staff Sergeant Strickland, a soldier of the 1st Armored Division "Old Ironsides". His equipment includes an M9 bayonet.

LAN-CAY M9 BAYONET

One Mission, Many Solutions, or the Eventful History of the M9 Bayonet

Lan-Cay was not the first company to produce the M9. The first M9 model, which replaced the M7 bayonet, was based on the Buckmaster 184. The American manufacturer Buck developed it further into the Model 188 and produced it for the California company Phrobis. When the contract for the M9 bayonet was rewritten in 1992, however, Lan-Cay underbid Buck-Phrobis and secured the first order for 30,500 bayonets, which was followed by one for 20,000 more. Since then Lan-Cay has been the largest producer of the M9 bayonet, although Ontario has also secured contracts in that time.

The M9 bayonet has an overall length of 30 centimeters, of which just under 17 centimeters is taken up by the blade (more than one centimeter more than its predecessors the M6 and M7). The cross guard is made of steel. Highly rustproof, modified 420 steel, roughly equivalent to 1.4034, is used for the blade. Almost six millimeters thick, it has a black oxide coating. More than half of the spine has a serrated edge, which can be used to cut not just wood, but laminated glass as well. Together with the sheath, which has a blade mounting/cutting plate screwed to its lower end, the Lan-Cay M9 can also be used as a wire cutter for hardened barbed wire.

The thermoplastic sheath has a spring clip inside for fixing the bayonet, and further security is provided by a cloth strap with snap. The sheath can be attached to the belt or equipment in a variety of ways. A plastic snap also allows it to be attached to and removed from equipment.

The M9 fits the M16 and M4 series assault rifles as well as the Mossberg 590 shotgun, which is used by the police and military under the name "Persuader". The Australian armed forces have also adopted the M9 for their version of the Steyr AUG assault rifle. Lan-Cay produces an M9-LW version. There are six round cutouts in the blade to reduce weight, the cross guard is made of aluminum. Both models are produced in a variety of colors by Lan-Cay.

There is also a sharpening stone on the back of the M9 sheath.

Type designation and manufacturer are stamped deep into the blade.

SPECIFICATION

Manufacturer:	Lan-Cay Inc.
Model:	M9 Bayonet
Blade Length:	183.0 mm
Blade Thickness:	6.0 mm
Blade Steel:	420HC
Overall Length:	313.0 mm
Weight (with/without sheath):	625/425 g
Handle:	Plastic
Sheath:	Plastic
Alternativen:	M9-LW
Manufactured in:	USA
Supply Number:	NSN 1095-01-227-1739

Between February and December 2002, Austria provided a contingent of seventy soldiers to serve with the NATO-led International Security Assistance Force (ISAF) in Afghanistan. The Austrians were based at Camp Warehouse in Kabul. Their duties included defending the camp against attack. Photographed in March 2002, this Austrian paratrooper was part of the crew manning a watchtower. He is armed with a 5.56-mm Sturmgewehr 77 and an Ontario M7 bayonet.

ONTARIO M7 BAYONET

A Classic in the History of Military Edged Weapons

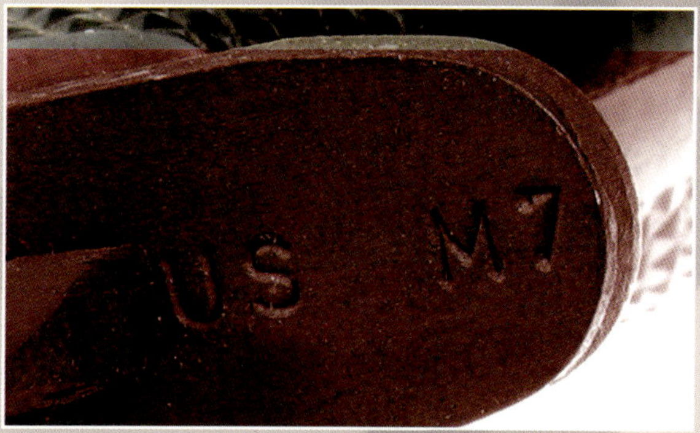

The type designation is stamped on the cross guard.

The M7 was originally developed for the Colt M16 assault rifle and was used by American troops during the Vietnam War. The original M7 (which was made in Germany by Eickhorn of Solingen as well as in the USA, Singapore, South Korea and the Philippines) replaced the M6 in 1964. The earlier design was largely retained as the M7, like the M6, was also intended for use as a fighting knife. The version seen in the photo on the left was made by Ontario and is almost identical to the original. Only the old M8A1 sheath with wire clip, which can be seen here, was replaced by a more modern version.

The center point blade is made of non-rustproof 1095 carbon steel with a black zinc phosphate coating to protect against rust and light reflection. The blade, which is 16.9 centimeters long and five millimeters thick, can also withstand heavy bending loads without breaking. The front half of the spine is ground. This bayonet is not suited for cutting barbed wire. The blade shape is based on the famous M3 fighting knife used by the American forces during the Second World War.

Just under 13 centimeters long, the voluminous plastic grip has a textured pattern over almost its entire surface. This design ensures a firm grip even in wet conditions. The cross guard, which is also black zinc phosphate-coated, not only sits firmly on the flash suppressor, but also protects the hand from slipping onto the blade.

The M7 fits a wide variety of assault rifles, including the Colt M16, M4, Diemaco C7, C8, Steyr AUG, IMI Galil AR, ARM, SS Artillery Regiment, Beretta AR70/90 and the FN FNC. It is a classic that will serve well into the 21st Century even though most armies are switching to multipurpose edged weapons which are both tools and weapons.

The sheath is a relic from the days of the Second World War.

SPECIFICATION

Manufacturer:	Ontario Knife Company
Model:	Bayonet M7
Blade Length:	173.0 mm
Blade Thickness:	5.0 mm
Blade Steel:	1095
Overall Length:	3030 mm
Weight (with/without sheath):	440/300 g
Handle:	Plastic
Sheath:	Plastic
Manufactured in:	USA
Supply Number:	NSN 1095-00-017-9701

25 December 2007, somewhere in the desert of western Iraq: Lance Corporal Deforest uses his OKC 3S bayonet to prepare a turkey for the grill. His comrades obtained the bird from some Bedouins a short time earlier in exchange for a few MRE (Meal Ready to Eat) rations and several bottles of drinking water.

ONTARIO OKC 3S MARINE CORPS BAYONET

Marine Bayonet with Special Buyer Qualities

A sharpening stone is located on the sheath below the belt loop.

The USMC emblem is molded into the handle.

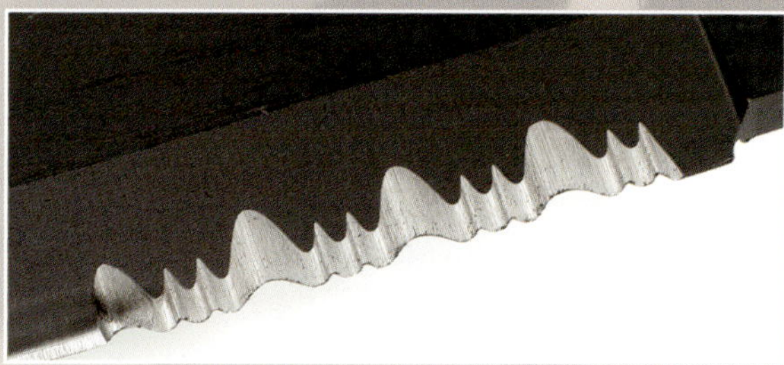

The serrated edge can be used to cut fibrous material.

In the last twenty years the Marines have tried out numerous bayonets. 5,000 examples of the M9 MPBS were purchased in 1991. In 2001 the Bajonett 2000 made by A. Eickhorn in Solingen was officially selected. The Marine Corps version was only slightly modified, one of the changes being the USMC-E/G/A emblem (Eagle/Globe/Anchor) engraved on the blade. Interestingly, there was no competition process, as the Americans were convinced that only Eickhorn could deliver the necessary quantities in the desired time. Several American competitors protested, however, and the contract was finally cancelled.

There was a new round of tests, in which various designs from seventeen manufacturers competed. The bayonets were tested by instructors at the Marine Corps Martial Arts Center of Excellence. Ultimately the Ontario Knife Company prevailed with its "Pattern 3" with serrated edge (OKC3S). Initial deliveries began on 13 December 2002 and by mid-2004 almost 100,000 examples had been delivered to the Marines.

The shape of the OKC 3S' handle and blade are very reminiscent of the USMC combat knife. The Bowie blade is 20.3 centimeters long, and the rear five centimeters of the main edge is serrated. The first nine centimeters of the blade spine are sharpened. Blade width is up to 3.5 centimeters and thickness is just under five millimeters. The cross guard and pommel are similar to those of the M7 bayonet.

The blade is made of an unspecified carbon steel which is selectively hardened (between 53 and 58 Rockwells) and given a zinc phosphate coating. The blade is marked on both sides, the bottom with COMBAT U.S.M.C., the top with OKC 3S ONTARIO KNIFE CO. There are also markings on both sides of the handle: the USMC title and the eagle-globe-anchor emblem. The 13.3-centimeter handle is made of Dynaflex plastic and is coyote brown in color. The thermoplastic sheath is also brown. The bayonet is held in place three ways: a spring in the sheath opening and two straps, which fit over the cross guard and handle. On the reverse side there is a ceramic-coated aluminum sharpener. Numerous attachment straps and loops enable the bayonet to be attached to clothing or equipment (including MOLLE).

SPECIFICATION

Manufacturer:	Ontario Knife Company
Model:	OKC 3S Marine Corps Bayonet
Blade Length:	203.0 mm
Blade Thickness:	5.0 mm
Blade Steel:	Carbon steel
Overall Length:	335.0 mm
Weight (with/without sheath):	600/380 g
Handle:	Dynaflex-Plastic
Sheath:	Thermoplastic Plastic
Manufactured in:	USA
Supply Number:	NSN 1095-01-596-3424

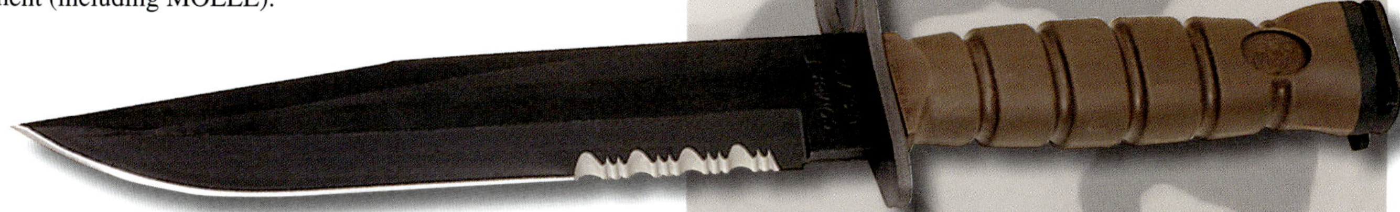

Afghanistan 2006: a DMR rifleman of the 10th Mountain Division scans the desert near Kandahar for Taliban fighters. Beside him is an M21/14 sniper rifle with a caliber of 7.62 x 51. By his right elbow is a Gerber MP600 multi-tool.
Photos: Sören Sünkler

TOOLS

Unfortunately, we have no detailed information about this photo. The man who took it has told us that the soldier on the right, Sergeant First Class White, used his VTAC to save the life of a fellow soldier during close combat.
Photos: American Tomahawk Company

AMERICAN TOMAHAWK COMPANY VTAC

In the Footsteps of Rogers' Rangers

The VTAC Tomahawk is impresses through its massive size.

It is amazing that in the age of modern warfare, with laser-guided bombs and unmanned drones capable of launching Hellfire missiles at pinpoint targets, archaic weapons such as axes and knives still have a place on the battlefield. This is exactly the way the American Tomahawk Company, founded by Peter LaGana in 1966, sees it. Some years ago, his legendary Tomahawk design for soldiers in Vietnam, more than 4,000 of which went to Southeast Asia, was again made available by Andy Prisco, the new head of the company.

Like the Vietnam original, the VTAC Model has an axe head of drop-forged 1060 carbon steel. As a result, the axe head's strength is higher than average. The head has a scratchproof epoxy powder coating to prevent corrosion. After the coating process, the edges are reworked by hand to provide the VTAC with the necessary sharpness.

The ATC company logo is found at the end of the VTAC's handle.

According to the manufacturer, the handle is made of ST synthetic material. ST stands for "Super Tough" and the material is most likely a fiberglass-reinforced plastic. The surface of the oval-shaped handle has a non-slip pattern similar to that seen on floor grates. At the end of the handle there is a lanyard hole.

The holster is jump-capable and there are numerous attachment options. It is also MOLLE-compatible. Two large Fastex buckles hold the VTAC in the sheath. Andy Prisco has finally achieved what Peter LaGana always wanted: the VTAC Tomahawk has officially been procured by the American military and has a NATO supply number.

SPECIFICATION

Manufacturer:	American Tomahawk Company, Inc.
Model:	VTAC
Blade Steel:	1060
Overall Length:	349.0 mm
Weight (with/without Sheath):	660/560 g
Handle:	SP Thermoplastic
Sheath:	Cordura
Manufactured in:	USA
Supply Number:	NSN 4210-01-518-7244

In addition to combat knives and tools, many American soldiers also carry simple belt cutters, so that they can cut themselves out of safety harnesses in an emergency. First Sergeant O'Connor of B Company has selected a Rescue Hook by Benchmade.

BENCHMADE MODEL 7 RESCUE HOOK

Specialized Emergency Tool for Cutting Through Equipment, Clothing and Boots

The NATO supply number appears beneath the company name.

The Rescue Hook can easily be grasped when wearing gloves.

There are situations in which it is imperative to free an injured man from belts, combat vest, belt, trousers and boots as quickly as possible – without endangering him with a sharp blade. The Rescue Hook was developed for exactly these kinds of situations.

Compared to other, similarly-shaped rescue tools, which often come with razorblade-like attachments, the Model 7 is made from a single piece of steel: 440C, with a Rockwell hardness of 58. The Rescue Hook has an overall length of 10.9 centimeters. The ergonomically-shaped handle is PVC-coated, which significantly enhances grip. It fits the hand well, even when wearing gloves. The 2.9-millimeter-thick spine has deep riffles just in front of the rounded end to provide a safe surface for the fingers. The end of the "blade" is rounded, and the guarded cutting edge curves sharply backward so that the Rescue Hook can be placed between skin and clothing and used safely.

The website www.7hook.com offers a demonstration of how a wounded soldier can be freed from his tactical vest, belt and a pant leg in just 15 seconds. Less than 20 seconds are required to cut bootstraps and the side of a boot from shaft to toe. The Model 7 should be seen as a supplement to a combat knife, for the Rescue Hook soon reaches its limitations against thick rope or other materials.

The MOLLE-compatible nylon sheath adds to the Rescue Hook's high functionality. It can also be attached to a belt or other items of equipment. A Velcro retention band with oversized pull-tab holds the Model 7 securely in the sheath. Three more models of the "Rescue Hook" family have since appeared: the smaller 5 Hook, the longer 8 Hook and the folding Houdini Pro rescue Tool with LED light and glass breaker.

SPECIFICATION

Manufacturer:	Benchmade Knife Company, Inc.
Model:	Model 7 Rescue Hook
Blade Length:	13.0 mm
Blade Thickness:	2.9 mm
Blade Steel:	440C
Overall Length:	109.0 mm
Weight (with/without Sheath):	80/45 g
Handle:	Vinyl plastic
Sheath:	Nylon
Options:	5 Hook, 8 Hook, Houdini Pro
Manufactured in:	USA
Supply Number:	NSN 4240-01-543-9618

The field rations issued to American troops are called Meals Ready to Eat (MRE). An MRE daily ration contains three single rations. This soldier, a member of the 1st Infantry Division "Big Red 1", is using his Gerber 600 tool to open the packet containing the main meal of his MRE,

GERBER MULTI-PLIER 600 NEEDLE-NOSE

Veteran Tool with Minor Shortcomings

Gerber has been in business for a long time. The company, which has produced such classics as the Mark II and the Gerber Gator Knife, was founded in Portland, Oregon in 1939. Today Gerber is part of the Finnish company Fiskars. Gerber has made multi-tools for many years, including the Multi-Plier 600 (here with needle nose), which has become a true tool classic.

What makes the MP 600 special is the ability to deploy the plier jaws with one hand. To do so, one presses the locking bolts inwards, allowing the plier jaws to be slid out of the handle halves. As innovative as this is, the tool is not perfect: the design allows for some movement of the plier jaws. As well, the plier grips cannot be opened unless the jaws are extended. As the tool blades can only be accessed with the handle open, the jaws must always be extended first.

The tool has two knife blades: one drop point and one serrated sheepsfoot blade. There are also three flathead screwdriver blades (one with a bottle opener), a Phillips head screwdriver blade, an awl, a file, a saw, a can opener and a lanyard hole. All blades lock in the open position.

A bit set is available to expand the tool's usefulness. In addition to a variety of bits, it includes a bit extender which attaches to the Phillips screwdriver. A sensible addition. With plier jaws extended the tool measures 16.8 centimeters. It weighs just under 230 grams. The multi-tool is delivered in a nylon belt pouch. It also forms part of the USAF survival vest worn by American air force pilots.

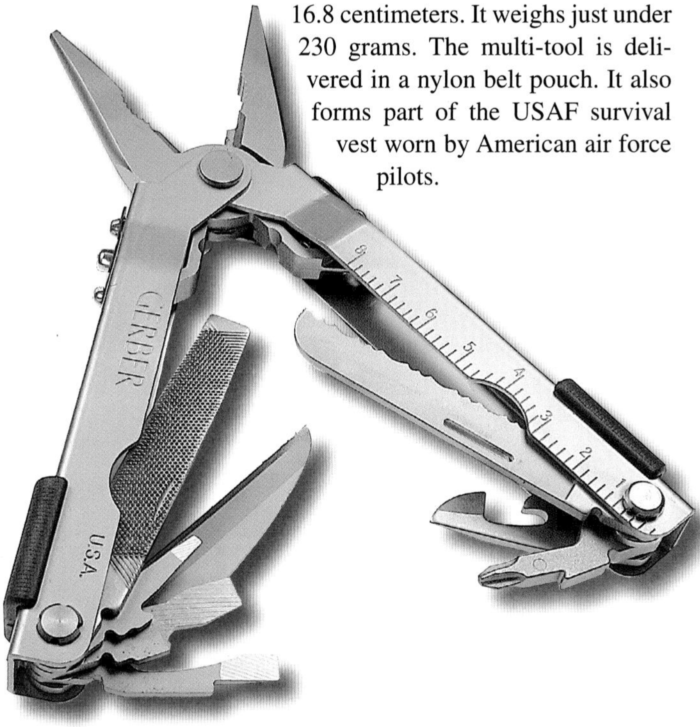

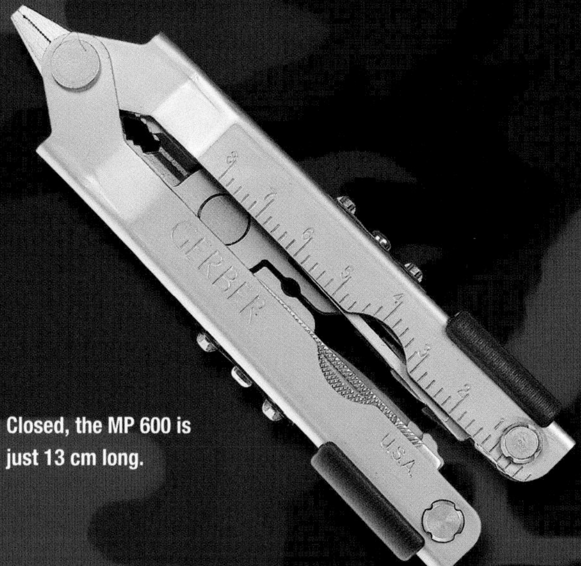

Closed, the MP 600 is just 13 cm long.

The tool is delivered with a classic Cordura holster.

SPECIFICATION

Manufacturer:	Gerber Legendary Blades
Model:	Multi-Plier 600
Blade Length (Wellenschliffklinge):	65.0 mm
Blade Thickness:	2.0 mm
Blade Steel:	—
Overall Length:	130.0/168.0 mm
Weight (with/without Sheath):	240/230 g
Handle:	Steel
Options:	Nylon
Alternativen:	Blunt-nose Pliers
Manufactured in:	USA
Supply Number:	NSN 5110-01-394-6249

January 2008: troops of the 4th Battalion, 9th Infantry Regiment of the 4th Stryker Brigade Combat Team of the 2nd Infantry Division patrol along the Diyala River northwest of Baqubah. This soldier is a machine-gunner armed with a 5.56-mm M249 SAW. He is carrying a Leatherman Charge TTi on the side of his ballistic vest.

LEATHERMAN CHARGE TTI

The Tool by Tim Leatherman

The tools made by the Leatherman Company have not missed any wars. For many years Leatherman has been an official supplier to the U.S. Army and the military forces of many other nations. Consequently, various versions have been assigned NATO supply numbers.

The Charge TTi model is one of the premium versions of the Charge series. The internal tools simply fold out and are protected by light, ergonomically-shaped titanium handle scales. For many years this was a weak point of the Leatherman PST and Supertool models, whose edges pressed uncomfortably into the palm when the tool was grasped firmly. To protect the user during use, the locking mechanism was considerably strengthened compared to the old Leatherman tools.

The two blades of high-grade S30V stainless steel have much superior edge-retention characteristics than the 420 steel usually used by Leatherman. The outside-accessible blades open easily and smoothly. The serrated blade also has a cutting hook at its tip.

The range of uses for the tools is almost endless. The equipment package includes needle-nose pliers, wire cutter, hard wire cutter, wood saw, shears, wood-metal file, diamond-coated file, large bit adapter, large screwdriver, small bit adapter, eight double-head bits (four hex, two Robertson, one straight edge, one Torx, two Phillips, a small Phillips and watchmaker's screwdrivers), ruler (inches/cm), can opener/bottle opener and a wire stripper. Nothing is left to be desired.

The titanium handle scales make the TTi highly valuable.

The holster can easily be attached to the MOLLE system.

SPECIFICATION

Manufacturer:	Leatherman Tool Group, Inc.
Model:	Charge TTi
Blade Length (Hauptklinge):	75.0 mm
Blade Thickness:	2.0 mm
Blade Steel:	S30V
Overall Length:	158.0/102.0 mm
Weight (with/without Sheath):	280/234 g
Handle:	Steel/Titanium
Sheath:	Cordura/Leather
Manufactured in:	USA
Supply Number:	None (privately procured)

Members of the 1st Battalion, 4th Marines in Fallujah with Eagle Talon axes. According to the manufacturer, a marine Lieutenant Colonel used his axe while searching houses. *Photo: RMJ Tactical*

RMJ TACTICAL KESTREL

Axe with Special Mission Profile

The spike on the axe head is capable of piercing Kevlar helmets.

Knifesmith Ryan Johnson has been part of the American knifemaking scene for years. His special field is axes. In 2001 Johnson met several members of the U.S. Special Forces at the Blade Show in Atlanta. They told him of an unusual idea: they wanted an axe in the style of the early American tomahawk capable of piercing a Kevlar helmet. Johnson accepted the challenge and designed the Eagle Talon model, which had a long handle to increase momentum and thus the axe's penetrative power. A roughly 8-centimeter-long spike on the back of the axe head took over the role of helmet breaker. To further increase depth of penetration, the spike was ground on four sides like a dagger, with the edges not sharpened.

To make the axe even easier to wield in action, Johnson developed a compact version with the name Kestrel. The Kestrel has an axe handle which is 33.5 centimeters long. The entire axe is made from a single piece of 4041 steel and possesses a thickness of nine millimeters over its entire length. All of the edges are hand-ground. The axe is selectively hardened to increase strength. The Kestrel has a dark-gray Mil-Spec phosphate coating to prevent rust. It nevertheless requires regular care with some oil.

550 Paracord winding or textured Micarta handle scales are both available. Both variants provide an outstanding grip. The sheath is made of Kydex and allows the axe to be carried ready to grip. Two nylon straps also secure the axe during parachute jumps or rappelling.

The Kestrel's Kydex sheath.

SPECIFICATION

Manufacturer:	RMJ Tactical
Model:	Kestrel
Blade Steel:	4041
Overall Length:	335.0 mm
Weight (with/without Sheath):	780/560 g
Handle:	Steel with 550 Paracord winding
Sheath:	Kydex
Manufactured in:	USA
Supply Number:	None (privately procured)

Unfortunately we have no detailed information concerning this photo provided by RMJ Tactical. The man in the photo is obviously a very satisfied customer, holding up greetings to RMJ from his armored vehicle. Right: This photo clearly shows how two axes are created in the forging process.
Photos: RMJ Tactical

RMJ TACTICAL SHRIKE

A Tough Forged Fighting Axe

Following the extremely successful introduction of his Eagle Talon and Kestrel axes, company founder Ryan Johnson again wanted to optimize the manufacture of his fighting axes. The result is the Shrike, which is drop forged instead of being laser-cut from a piece of material. The forging process always creates two axes at a time. After the forging blank has cooled, the excess material is removed and the blank is further worked by hand.

As with his other models, Johnson relies on 4140 chrome-molybdenum steel, which has a low Rockwell hardness of about 30 in the area of the handle. Hardness reaches 55 Rockwells in the area of the edges. The surfaces have a Mil-Spec phosphate coating.

Like the Eagle Talon and the Kestrel, the Shrike has a spike-like rear part, which is capable of piercing Kevlar helmets, wood and even car doors. The handle has an injection-molded grip of "Super Tuff" nylon. This procedure results in an electrically-insulated grip and a better textured surface. Unlike the other RMJ axes, the Shrike has an oval-shaped handle, which fits comfortably in the hand. The buyer can choose a black or sand-colored handle.

Johnson has come up with an additional function for the handle: like on a survival knife with hollow handle, he has concealed a fine-grained sharpening stone in the handle. Unscrewing the butt cap reveals the stone, and the end of the threaded section is shaped so that it can function as a striking enhancer. The end of the handle can thus be used to break glass.

RMJ Tactical provides a Kydex quick-draw sheath which attaches to the equipment. There is also an adapter for attaching the sheath to the MOLLE system.

The sharpener is an integral part of the axe. It inserts into the handle.

Each axe bears a unique identification number.

SPECIFICATION

Manufacturer:	RMJ Tactical
Model:	Shrike
Blade Steel:	4041
Overall Length:	394.0 mm
Weight (with/without Sheath):	870/652 g
Handle:	ST801-Plastic
Sheath:	Kydex
Manufactured in:	USA
Supply Number:	None (privately procured)

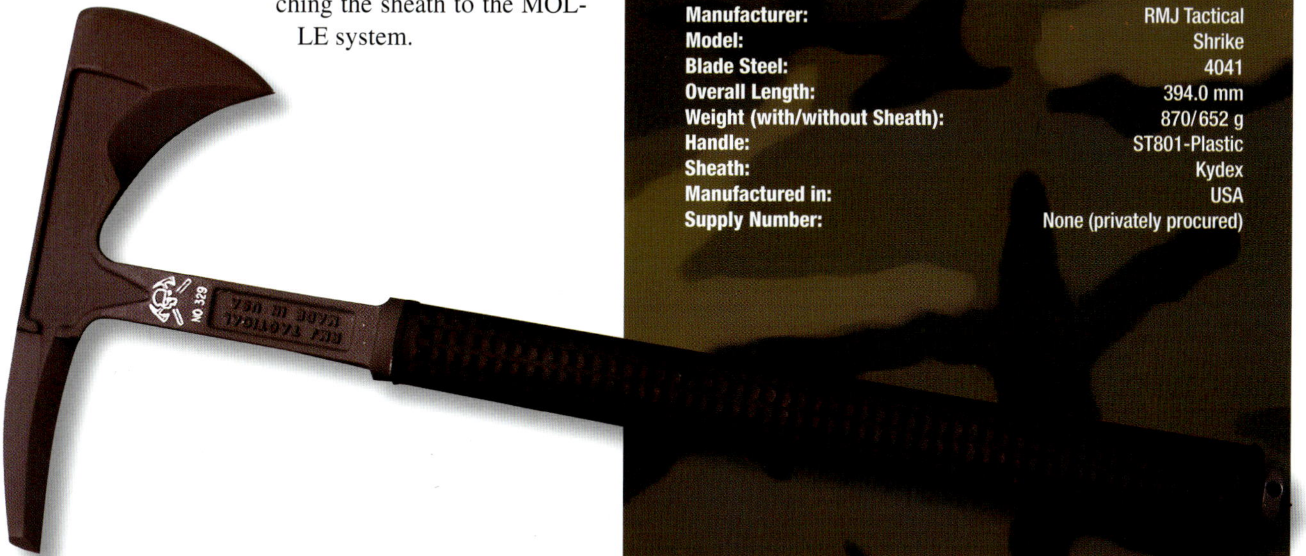

The equipment carried on the survival vests of U.S. Marine Corps pilots includes an SOG B60 multifunction tool. The marine in the photo is the pilot of a two-seat F/A-18D Hornet of Marine Fighter Attack Squadron (All Weather) 225 "Vikings". At the turn of the year 2007-2008, the squadron was operating from Al-Asad airbase in western Iraq.

SOG POWER LOCK BLACK OXIDE

All-Rounder from a Proven Manufacturer

Main blade, awl, screwdriver and bottle opener.

A multi-tool is often more useful than a large, fixed military knife – at least for everyday chores. The American company SOG, known for its military knives, also has a great deal of experience in making multi-tools. After Leatherman, it was the second manufacturer to offer folding pliers with extra tools.

The PowerLock is made of 1.4034 steel which, while not particularly hard, is extremely rustproof. The lockable tool blades are only accessible when the handles are opened. The tool recesses are provided with a removable cover with rounded corners. The hardness of the tools varies. The PowerLock B60 has needle-nose pliers and a crimper device within the confines of the handle. The tool also has a hollow-ground blade with partial serrated edge (Rockwell hardness of 52, 1.8 mm thickness), a metal file with three cutting surfaces, plus a wood saw, an awl, scissors, several screwdrivers hardened to 54 HRC (two slotted, one Phillips) and a 1/4 inch tool adapter which accepts 1/4 inch sockets and, with an additional adapter, all 1/4 inch hex drive bits.

Phillips screwdriver, can opener and saw.

A special feature of many SOG tools is the Compound Leverage System. The plier jaws open and close over a total of four gears (two per handle scale) which are part of the handle. Necessary hand pressure is reduced by up to 50 percent. If both hands are not free to open the two handle scales and extend the plier jaws, the pliers can be opened one handed with a flick of the wrist.

The PowerLock has a black oxide coating and can be completely disassembled using a 1/4 inch hex key. As all of the tool blades can be purchased separately, one can simply replace worn components. The tool is delivered with a sturdy belt sheath made of cowhide. Anylon sheath is also available, equipped with a plastic clip.

SPECIFICATION

Manufacturer:	SOG Specialty Knives & Tools, LCC
Model:	PowerLock B60
Blade Length (Hauptklinge):	67.0 mm
Blade Thickness:	1.8 mm
Blade Steel:	1.4034
Overall Length:	119.0/179.0 mm
Weight (with/without Sheath):	–/257 g
Handle:	1.4034
Sheath:	Leather
Options:	Uncoated, with C4 spike or V cutter
Manufactured in:	USA
Supply Number:	None (privately procured)

S0529
SGT SOLIS
AIRBORNE

GLOSSARY: KNIFE TERMINOLOGY

ABS
A composite plastic which is highly shock- and heat-resistant as well as machine washable.

Aluminum 6061
Often called aircraft aluminum (ISO-Norm:AlMg1SiCu). It is extremely corrosion-resistant. It is preferred for knife handles on account of its low weight and relatively high resilience.

Arc Lock
Locking system developed by the SOG Company, with a locking pin mounted on a movable lever.

Assisted Opener
A semi-automatic spring mechanism on a folding knife. The blade must be opened slightly by hand before a spring takes over and moves the blade into the full-open position.

Back-Lock
Also Lock-Back. By far the most widely-used locking system seen on folding knives. This design functions with a locking arm along the handle spine being shaped with a hook or post end that fits, under spring pressure, into a notch on the back of the blade behind the pivot. To release, one must press on the rear end of the hook, thus lifting it forward. There are versions with long or short hooks. The best known folding knife with the Back-Lock system is the Buck Folding Hunter, which was first introduced in 1964.

Bead Blast Finish
A smooth matte finish on knife blades and other components produced by glass bead blasting. The bead streams are often exchanged for sand streams, which produces a coarser surface finish. The bead streams produce small craters and irregularities, consequently this process is used mainly on more expensive knives.

Blackening
A procedure for blackening iron-based materials such as cast iron, forged iron, unalloyed or alloyed steel. The object is placed in a bath of caustic soda, sodium nitrite / nitrate, wetting agents and stabilizers, creating a black iron oxide finish whose black tone depends on the quality of the materials, surface preparation and the type and condition of the bath. The procedure is used to give the object a pleasing appearance and increase its resistance to corrosion.

Clip
As a rule, a metal strip or loop mounted on the handle of a folding knife. The clip allows the knife to be stuck in a pocket or on a belt. This solution was first introduced by Spyderco. There are also sheaths for fixed knives equipped with a clip.

Concealex
Similar to Kydex in material and characteristics, used by the manufacturer Cold Steel.

Cordura
A weave of nylon fibers which, like many other synthetic materials, originates from DuPont. Melting point is 210 degrees. The nylon fibers are cut, respun and then woven. The result is a material that is extremely tough and resistant to abrasions, which is also used in the outdoors and motorcycle fields.

Cross Guard
See Guard

Elastomer
Elastomers are synthetic materials, such as polyurethane, that hold a shape but can be reshaped. A feature of the elastomer group is that it can be stretched and bent with little effort and then returns to its original shape. They are micro-molecules held together only by a few chemical cross-linkages.

False Edge
Ground, but not sharp ground area on the blade spine. As a rule it begins at the tip and extends several centimeters. The result is a finer tip, making the blade better suited for stabbing.

Flipper
The flipper is a ‚trigger' for flipping a knife open. In the closed position, the flipper appears as a protrusion from the back of the handle where it can be pushed, causing the knife to open. With the blade open, the flipper protects the hand. The flipper was first used by American knife maker Kit Carson.

Forprene
A rubber-like Vulcanized Thermoplastic Elastormer used by Extrema Ratio for knife handles.

Frame Lock
A frame lock system is similar to a liner lock, however the knife with a frame lock typically does not have handle scales but rather is a simple frame design in which the two sides of the handle are, alone, the basis of both the frame and handle. The two sides of the handle form the frame and are often made from steel cut thicker than what any traditional liner lock would be. It is this thickness from which the frame lock draws its considerable strength. In other respects, the frame lock functions very much like a liner lock. A portion of the frame which makes up one side of the knife handle is cut and given a set so it angles inward toward the interior of the handle with the tip engaging the bottom of the blade which has been cut at an angle like a ramp. It is released by applying pressure to move the piece of frame away from the blade bottom and back toward its natural side and is held in place there by closing the bla-

de. Upon opening, the tension the lock piece has been given causes it to snap across the bottom of the blade engaging at its furthest point of travel as determined by the angle cut into the blade bottom. Also known as Integral Lock.

FRN
Abbreviation for Fiberglass Reinforced Nylon, a light and tough synthetic material.

Full Tang
See Tang

G-10
An extremely durable makeup of layers of fiberglass soaked in resin, then highly compressed and baked. Impervious to moisture or liquid and physically stable under climate change. The material is usually sandblasted, resulting in a rough surface which provides a good grip.

Guard
Element between the blade and handle that protects the hand. The guard can either be on both sides of the knife (cross guard) or only on the bottom.

Hand Guard
See Guard

Integral Lock
See Frame Lock

Integral Knife
Term for a fixed knife in which the blade, tang and cross guard are made from a single piece of metal (milled or forged).

Kalgard
English company specializing in coatings and lubricants. Manufacturer of polymer-based coatings for knife blades.

Kevlar
An organic synthetic fiber developed at DuPont in 1965. Fibers of Kevlar consist of long molecular chains produced from poly-paraphenylene terephthalamide. The chains are highly oriented with strong interchain bonding, which results in a unique combination of properties. Features of Kevlar fibers include high tensile strength at low weight, low elongation to break high modulus (structural rigidity), low electrical conductivity, high chemical resistance, low thermal shrinkage, high toughness (Work-To-Break), excellent dimensional stability, high cut resistance, flame resistant, self-extinguishing. Kevlar has become well known because of its use in bullet-proof vests.

Kraton
A thermoplastic elastomer offering excellent flexibility and memory elasticity even under harsh conditions. It is over-molded onto the

This soldier used his Gerber A-F Combat Folder to remove a poster which, in addition to bearing the image of Muqtada al-Sadr, calls for armed action against the coalition forces and other religious groups in Iraq.

handle for added grip surface and tactile qualities.

Kydex

A mixture of acrylic and PVC, which offers outstanding impact and chemical resistance as well as fire retardancy. Kydex is light, scratch- and shock-resistant, chemically-resistant and is ideally suited for making knife sheaths.

LAWKS

"Lake And Walker Knife Safety", a technical innovation by CRKT. The original LAWKS mechanism consists of a switch on the handle of the knife, connected to the liner lock. When pulled back, the liner lock can easily be pushed back to close the knife. But when the safety switch is pushed forward, it acts as an extra lock ensuring the liner lock can not be used to close the knife, making the folding knife a virtual fixed-blade tool.
The AutoLAWKS mechanism was designed after professional military/law enforcement consultants pointed out that sudden tactical or immediate operation situations would call for focus on something other than actuating a safety switch under duress. The AutoLAWKS automatically activates, as the knife is opened. To close the blade again, the user has to pull the switch back to free the liner lock like the original LAWKS, but the AutoLAWKS switch automatically returns to a neutral position

Liner

The liner is a thin piece of metal, usually steel or titanium, that is found under the handle scales. It serves as reinforcement of the handle and is also where the lock on a liner lock knife comes from.

Liner Lock

A locking mechanism with origins in the area of handmade Custom Knives which was developed into its present form by American knife-maker Michael Walker.
In its most common configuration, the liner lock is created from the steel liners most knives incorporate. Typically cut from the liner on the reverse side of the knife (when looking at knife, blade tip pointing left and edge facing up), a section of the liner on this side is bent away and angled inward toward the interior of the knife. This section of liner bending inward is given a natural „set" in this shape so its „at ease" position is angling away from the liner from which it is cut and cutting across the interior section of the knife handle. The liner now can only go back to its old position with manual force.
The end of the liner lock which is closest to the blade is cut so as to engage with the bottom end of the blade below the pivot. This bottom end of the blade is cut at an angle so the liner moves across the blade bottom to the point at which the angle stops its movement (ideally in the middle of the blade). The blade is now said to be locked in the open position. To disengage the lock, one must manually move the liner to the side, away from the blade bottom before closing the blade. Once closed, the presence of the blade prevents the liner from moving, but when the blade is opened, the tension in the steel snaps the liner open, stopping at the point where the bottom end of the blade angle prevents further travel. The blade is now open and locked again.

Lock Back

See Back Lock

Micarta

Layers of linen cloths soaked in resin and pressured together. It is a very popular material for making knife handles, as it is easy to work. Most commonly used on more stylish knives due to its rich color finish qualities.

MOLLE

An acronym for Modular Lightweight Load-carrying Equipment. It is used to define the current generation of load-bearing equipment and rucksacks utilized by the United States armed forces, especially the United States Army.

Open Frame

A folding knife handle constructed like a framework. The spine is open and the handle scales are joined by spacers or screws. This design makes it easy to clean the knife.

Paracord

Trade name for a strong nylon cord Parachute cord (also paracord or 550 cord) is a lightweight nylon kernmantle rope originally used in the suspension lines of US parachutes during World War II. Once in the field, paratroopers found this cord useful for many other tasks. It is now used as a general purpose utility cord by both military personnel and civilians. The braided sheath has a high number of interwoven strands for its size, giving it a relatively smooth texture. The all nylon construction makes paracord fairly elastic; depending on the application this can be either an asset or a liability.

Quillion

The projections that keep the hand from sliding forward onto the cutting edge, or backward off the butt of the knife handle. A knife may have a front quillon or quillons, and rear quillon or quillons.

Ricasso

The ricasso is the unbevelled portion of the blade after the blade grind where the blade meets the handle. Usually a maker's mark can be found here.

Santoprene

A polyurethane synthetic material used in the manufacture of knife handles. It has similar levels of flexibility and durability as rubber but has a longer life in extreme environments.

SAT

SOG Assisted Technology. SOG's designation for the company's semi-automatic opening system.

Satin Finish

High-grade finish achieved with a polishing machine. It appears as a regular structure of extremely fine grind marks, which as a rule run at right angles to the longitudinal axis of the blade. Various levels of quality are determined by the fineness of the polish.

Secure-Ex

Cold Steel company designation for a synthetic material used to make knife sheaths.

Spring Knife

A folding knife whose blade springs from the handle at the touch of a button or lever. There are various types of spring knife, using spiral or leaf springs.

Tang

An extension of a blade that extends into the handle. There are flat tangs and round tangs. The first has handle scales (sides) attached on both sides, while the second has the handle stuck onto the tang and fixed with screws, rivets or glue. Most high-grade knives have a full tang, which extends the full length of the handle.

Tek-Lok

Product name for a sheath mounting system which allows a Kydex sheath to be attached to a belt or equipment in a variety of positions.

Thermoplastic

A thermoplastic (sometimes written as thermo plastic) is a type of plastic made from polymer resins that becomes a homogenized liquid when heated and hard when cooled. When frozen, however, a thermoplastic becomes glass-like and subject to fracture. These characteristics, which lend the material its name, are reversible. That is, it can be reheated, reshaped, and frozen repeatedly.

Thermorun

A rubber-like thermoplastic material made by Mitsubishi Chemicals. Cold Steel and Fällkniven use it for knife handles (the material is very similar to Kraton by DuPont).

Thumb Stud

Opening aid designed to enable one-handed opening with the thumb. Can be installed on one or both sides of the knife – for right- and left-handed users.

Tip Up / Tip Down

Two ways of carrying a folding knife – with the blade tip up or down. How the knife is carried depends on how the attachment clip is mounted on the knife. On many knives the clip is adjustable for this purpose.

Titanium

A light, extremely strong, completely rust-free, antimagnetic and non-allergic material. Its most common application in the manufacturing of knives is for handle liners and scales, and in rare cases as blade material. Pure titanium is generally used for scales and handles, while for liners (especially on Liner Lock knives) the spring-hard titanium alloy 6Al4V is used. Titanium is lighter than steel by about one-third but possesses the same tensile strength.

Zytel

Trade name for a lightweight glass filled nylon engineered for strength and durability. Zytel is heat and cold resistant, strong and non-abrasive. During manufacture it can be sprayed directly onto the tang, and is therefore widely used in mass production.

HOOPS

ADDRESSES

All knife brands in alphabetical order and a list of suppliers

American Tomahawk: Modi's Knives and Gear
Benchmade: ACMA, Herbertz, Specialist shops
Cold Steel: Herbertz, Specialist shops
CRKT: Specialist shops (Switzerland)
Eickhorn-Solingen Ltd.: Eickhorn, Herbertz, Toolshop, Specialist shops
Extrema Ratio: Messer Barthel, Katachi-Art, Ninjutsu-Akademie, Wolfster
Gerber: Fulmatech, Herbertz, Specialist shops
IschMasch: sale to authorities only
Ka-Bar: Herbertz, Specialist shops
Kershaw: Specialist shops, Kai, Linder, Messer Matheisen, Katachi-Art, Wolfster, Toolshop, Ninjutsu-Akad.
Lan-Cay: sale to authorities only
Leatherman: S. Hintz, Herbertz, Specialist shops
Lile Knives: no longer exists
Marto-Phrobis: no longer exists
Mikov: Sbimex
MOD: Herbertz
M-Tech: Ninjutsu-Akademie, Wolfster (Switzerland: Klötzli)
Nemex: Herbertz, Toolshop, Wolfster, Ninjutsu-Akad., Katachi-Art, Specialist shops
Pohl Force: Pohl Force
Randall: Nohl
RMJ Tactical: direct sale only in USA
Smith & Wesson: Herbertz, Hebsacker, Rudolf Weber jr., Specialist shops
SOG: Fulmatech, Specialist shops
United Cutlery: Swords & More
Vox Knives: Vox Knives
XS Scuba: direct sale only in USA

SOURCE OF SUPPLY/CONTACT INFORMATION

ACMA Reus
www.acma.de
Tel. 09367-981616

Eickhorn-Solingen Ltd.
www.original-eickhorn.de
Tel. 0212-66050

Extrema Ratio
www.extremaratioknivesdivision.eu
infodealers@extremaratio.com

Siegfried Hintz
www.s-hintz.de
Tel. 0611-18689-0

Fulmatech
www.fulmatech.de
0931-46584985

Hebsacker
www.stahlwarenhaus-hebsacker.de
Tel. 0791-84091

Herbertz
www-herbertz-messerclub.de
Tel. 0212-201061

Kai Europe
www.kai-europe.com
Tel. 0212-23238-0

Katachi Art & Bladeshop
www.katachi-art.de
Tel. 040-4601448

Klötzli
www.klotzli.com
Tel. +41-(0)34-4222378

Linder, Carl Nachf.
www.linder.de
Tel. 0212-330856

Messer Barthel
www.messer.barthel.cc
Tel. 02203-9803888

Messer Matheisen
www.messer-matheisen.de
Tel. 0221-2580494

Mikov
www.mikov.cz
Tel. +420-412-394102

Modi's Knives and Gear
www.modisknivesandgear.de
Tel. 02234-697181

Ninjutsu-Akademie
www.knifetom.net
Tel. 07131-772787 (15-17 h)

Nohl, Messerschmiede
www.nohl-giessen.de
Tel. 0641-970787

Pohl Force
www.pohlforce.de
Tel. 02174-8966735

Rudolf Weber jr.
www.weberknives.de
Tel. 0212-59213

Sbimex GmbH
www.sbimex.de
Tel. 0375-27213926

Swords & More
www.swords-and-more.de
Tel. 040-97071570

Toolshop
www.toolshop.de
Tel. 03303-2174848

Vox Knives
www.voxknives.com
Tel. +45-616-58664

Wolfster's Knife Shop
www.wolfster.de
Tel. 06073-742379

Photo: Sören Sünkler

A fighting axe by the American knife maker Daniel Wnkler, developed by him for Marine special units. Beside it is a Pohl Force Alpha One prototype folding knife with digital camouflage.

ACKNOWLEDGEMENTS

The authors wish to thank the following persons and companies for their assistance in creating this book:

Persons

Bremer, Rod
Brüggendick, Dirk
Erber, Dan
Eickhorn, Jörg
Elishewitz, Allen
Fleitmann, Thomas
Grabosch-Pohl, Nicole
Held, Michael
Hoch, Michael
Hunt, Alicia
Idel, Carsten
Laible, Thomas
Lang, Oliver
Leckebusch, Tobias
Johnson, Ryan
M., Tobias
Milliken, Paul
Mück, Helmut
Nohl, Thomas
Prisco, Andy
Randall, Gary
Sänn, Jens
Schmitz, Wilhelm
Sünkler, Sören
Tigges, Jens
Voxknaes, Jesper
Weiler, Gernod
Wieland, Hans Joachim
Winkler, Daniel
Wydeau, Caroline
sowie den Mitgliedern des Tactical Forums (www.tacticalforum.de)

Companies

Benchmade (www.benchmade.com)
C. Jul. Herbertz (www.herbertz-messerclub.de)
Columbia River Knives and Tool (www.crkt.com)
Fulmatech (www.fulmatech.de)
Gerber Knives (www.gerberstore.com)
Group H3 (www.grouph3.com)
Gurkha Imports (www. gurkha-imports.com)
Haller Stahlwarenhaus (www.haller-stahlwaren.de)
Ka-Bar Knives (www.ka-bar.com)
K-ISOM (www.k-isom.com)
Modis Knives and Gear (www.modisknivesandgear.com)
Outdoor Marketing International (www.outdoor-mi.com)
Randall Knives (www.randallknives.com)
RMJ Tactical (www.rmjtactical.com)
Sbimex (www.sbimex.de)
Venatus Verlag (www.venatus.de)
Vogt Schild Deutschland, Redaktion VISIER (www.visier.de)
Vox Knives (www.voxknives.com)
Waffentechnik Borkott & Eickhorn (www.waffentechnik-solingen.de)
The Original Eickhorn-Solingen Co. Ltd. (www.original-eickhorn.de)

ABOUT THE AUTHORS

Dietmar Pohl

Born in Leverkusen in 1966, for 25 years Dietmar Pohl has been an enthusiastic collector of knives. He has designed more than 60 different knife models. He has also written three books on the topic of knives, which have been translated into several languages, as well as numerous magazine articles. For 14 years Dietmar Pohl held a senior position in the Solingen knife industry..

Dietmar Pohl's passion for knives has influenced him almost his whole life long. After studying economics, in 1994 his hobby became his career when he became a head of marketing, designer and project manager in the Solingen knife industry. He realized his ideas and designs with engagement, competence, extensive knowledge and a sense for trends, forms, technology and materials. His special focus lay in the field of modern EDC knives and tactical knives for the military and police.

Dietmar Pohl himself was never a member of a military or police special unit. His extensive knowledge is based on personal contacts, conversations and confidential exchanges of experience with members of GSG9, SEK, combat divers, KSK, paratroopers, SWAT, Marines Force Recon, London MET Police and other units and authorities. He takes every opportunity to gain first hand experience in order to expand his practical knowledge: Bundeswehr defense exercises for civilian executives held by the air force and army, a SWAT course in Sterling, Virginia, bayonet and knife combat training with the U.S. Marines Force Recon at Camp Pendleton, San Diego, and more.

Since 2008 Pohl has been realizing his knife designs under his own name. With the company POHL FORCE (www.pohlforce.de) he can realize his ideas 100 percent and pat attention to every detail without having to consider pre-existing restrictions.

Carl Schulze

Carl Schulze, a photo journalist specializing in military subjects and well known within specialist circles, was born in Springe near Hanover in 1968. He became interested in military topics at an early age. Growing up during the Cold War, as a teenager he spent his free time photographing the last major NATO exercises in the 1980s..

Between 1986 and 1994 Carl then served as a soldier in the Bundeswehr. For several years he was stationed in Munster, where he was a basic training instructor. At the beginning of the 1990s he began publishing his photos in military magazines, and his first article was published in the late summer of 1990. Since then Carl Schulze has written hundreds of articles for journals in and outside Germany, including Y, the official magazine of the Bundeswehr, Combat & Survival, RAIDS, Assault, Special Ops Magazine, Schweizer Soldat, Deutsches Waffenjournal, Visier, Kommando International Special Operations Magazine and Assault Journal.

Since leaving the Bundeswehr, Carl Schulze has worked primarily as a photo journalist. He regularly travels to the crisis and war zones of the world. Until 1998 he frequently reported on operations by the UNPROFOR, IFPR and SFOR troops in Bosnia-Herzegovina. In June 1999 NATO marched into Kosovo, and several days later Carl Schulze followed to report on the troops taking part in the KFOR operation.

Since 2002 Carl Schulze has regularly accompanied American and ISAF troops in Afghanistan. Since 2003 he has also reported from Iraq. By 2009 Carl had visited this crisis and unrest-torn country ten times, reporting on operations by units of the U.S. Army, the USMC and British forces. Carl is also involved in military themes n his spare time. His hobbies include military history and reenactment.

Schulze has published numerous photo books on a variety of military topics, for example on IFOR, the Warrior armored troop carrier, the M113 APC, the German paratroops, the AMF(L), the KFOR, the Stryker IAV fighting vehicle, Gun Trucks in Iraq, the 1st (U.S.) Cavalry Division in Iraq, the HEMTT Truck family, the Land Rover 110 Defender, the 24th Airmobile Brigade and the Bv 206 Husky.

GRENADES
BATTLE OVERRIDE
MODE
POWER
UPPER
LOWER
FIRE
OFF
CHG
MENU ON/OFF
RTCL SEL
LEVEL
OUT
ZOOM
IN
DAY NIGHT
RANGE
SPEED
LAMP TEST
BRIGHTNESS
ZELL